T0185222

Synthesis Lectures on Digital Circuits & Systems

Series Editor

Mitchell A. Thornton, Southern Methodist University, Dallas, USA

This series includes titles of interest to students, professionals, and researchers in the area of design and analysis of digital circuits and systems. Each Lecture is self-contained and focuses on the background information required to understand the subject matter and practical case studies that illustrate applications. The format of a Lecture is structured such that each will be devoted to a specific topic in digital circuits and systems rather than a larger overview of several topics such as that found in a comprehensive handbook. The Lectures cover both well-established areas as well as newly developed or emerging material in digital circuits and systems design and analysis.

Tyler Kerr · Steven Barrett

Arduino IV: DIY Robots

3D Printing, Instrumentation, and Control

 Springer

Tyler Kerr
Innovation Wyrkshop
University of Wyoming
Laramie, WY, USA

Steven Barrett
Electrical and Computer Engineering
University of Wyoming
Laramie, WY, USA

ISSN 1932-3166   ISSN 1932-3174 (electronic)
Synthesis Lectures on Digital Circuits & Systems
ISBN 978-3-031-11211-9   ISBN 978-3-031-11209-6 (eBook)
https://doi.org/10.1007/978-3-031-11209-6

© The Editor(s) (if applicable) and The Author(s), under exclusive license to Springer Nature Switzerland AG 2022

This work is subject to copyright. All rights are solely and exclusively licensed by the Publisher, whether the whole or part of the material is concerned, specifically the rights of translation, reprinting, reuse of illustrations, recitation, broadcasting, reproduction on microfilms or in any other physical way, and transmission or information storage and retrieval, electronic adaptation, computer software, or by similar or dissimilar methodology now known or hereafter developed.

The use of general descriptive names, registered names, trademarks, service marks, etc. in this publication does not imply, even in the absence of a specific statement, that such names are exempt from the relevant protective laws and regulations and therefore free for general use.

The publisher, the authors, and the editors are safe to assume that the advice and information in this book are believed to be true and accurate at the date of publication. Neither the publisher nor the authors or the editors give a warranty, expressed or implied, with respect to the material contained herein or for any errors or omissions that may have been made. The publisher remains neutral with regard to jurisdictional claims in published maps and institutional affiliations.

This Springer imprint is published by the registered company Springer Nature Switzerland AG
The registered company address is: Gewerbestrasse 11, 6330 Cham, Switzerland

Preface

This book is about the Arduino microcontroller and the Arduino concept. The visionary Arduino team of Massimo Banzi, David Cuartielles, Tom Igoe, Gianluca Martino, and David Mellis launched a new innovation in microcontroller hardware in 2005, the concept of open-source hardware. Their approach was to openly share details of microcontroller-based hardware design platforms to stimulate the sharing of ideas and promote innovation. This concept has been popular in the software world for many years.

This book, *Arduino IV: DIY Robotics—3D Printing, Instrumentation, and Control*, is about designing and building your own robots. As with other volumes in the series, we begin with a quick overview of the Arduino Integrated Development Environment (IDE) used to write sketches, and the hardware systems aboard the Arduino UNO R3 and the Mega 2560 Rev 3. We then examine concepts common to all robots. Along the way, we explore off-the-shelf robots: wheeled robots, tracked robots, and also a robotic arm. After the Arduino IDE and hardware introduction, we launch into "do it yourself" or DIY concepts. We start with an introduction to low-cost 3D printing. These concepts will allow you to design and print your own custom robot parts and chassis. We then explore concepts to sense a robot's environment, move the robot about, and provide a portable power source. We conclude with several DIY robot projects.

Approach of The Book

This book is part of a multi-volume introduction to the Arduino line of processors. The book series also serves as the "fourth edition" of "Arduino Microcontroller Processing for Everyone!" When discussing plans for a fourth edition, Joel Claypool and I (sfb) decided to break the large volume up into a series of smaller volumes to better serve the needs and interests of our readers. I have tried to strike a balance between each volume being independent of one another while holding to a minimum of information contained in other volumes. For completeness and independence, this volume contains tutorial information on getting started, microcontroller interface information, and motor control partially contained in some of the other volumes and related works completed for Morgan and

Claypool. I have identified via chapter footnotes the source of this information contained elsewhere in the series.

This book is divided into a series of chapters to prepare you to design and build your own robots. The concepts provided may be used for small or large robots.

As with other volumes in the series, Chap. 1 begins with a quick overview of the Arduino Integrated Development Environment (IDE) used to write sketches. We also explore hardware systems aboard the Arduino UNO R3 and the Mega 2560 Rev 3. The chapter concludes by exploring two off-the-shelf robots: the Dagu Rover 5 robot and the Tinkerkit Braccio Arduino robotic arm. We equip robots with additional features throughout the book.

Central to our discussion of robots is the ability to design and fabricate custom robot parts and chassis. Chapter 2 provides instruction on low-cost 3D printing to allow custom fabrication of robot components.

In the next three chapters, we learn how to equip robots with sensors, a drive train, and a power source. Chapter 3 explores concepts to allow a robot to sense its location, avoid obstacles, and sense robot status and the surrounding environment.

Chapter 4 explores different types of motors and actuators used aboard and to drive robots. We explore in detail how to interface and control high-power motors from a low-power microcontroller.

Chapter 5 explores portable power sources for different robots. We explore different battery types and how to size a battery system for a specific application.

In Chap. 6, we apply the book concepts to the design of a 4WD robot.

Acknowledgments

A number of people have made this book possible. I (sfb) would like to thank Massimo Banzi of the Arduino design team for his support and encouragement in writing the first edition of this book: "Arduino Microcontroller: Processing for Everyone!" We would also like to acknowledge Joel Claypool of Morgan and Claypool Publishers. Joel has provided his publishing expertise and support to a number of writing projects. He has provided many novice writers the opportunity to become published authors. His vision and expertise in the publishing world made this book possible. We would also like to thank Dharaneeswaran Sundaramurthy of Total Service Books Production for his expertise in converting our final draft into a finished product.

We also thank Adafruit, DFRobot, Microchip, Mikroe, SparkFun, and Traxxas and for their permission to use images of their products and copyrighted material throughout the text series. Several Microchip acknowledgments are in order:

- This book contains copyrighted material of Microchip Technology Incorporated replicated with permission. All rights reserved. No further replications may be made without Microchip Technology Inc's prior written consent.
- *Arduino IV: DIY Robots—3D Printing, Instrumentation, and Control* is an independent publication and is not affiliated with, nor has it been authorized, sponsored, or otherwise approved by Microchip.

Laramie, WY, USA   Tyler Kerr

2022   Steven Barrett

Contents

About the Authors

Tyler Kerr M. S. received a B.A. in Geoscience from Franklin & Marshall College in Pennsylvania in 2011, and an M.S. in Geology (Paleontology) from the University of Wyoming in 2017. His background in paleontology and interest in emergent technology led him to a career in 3D printing, 3D scanning, digital rendering, and digitizing museum collections. Today, Kerr manages the Innovation Wyrkshop makerspace, one of the largest academic makerspaces in the Mountain West. In addition to the Innovation Wyrkshop, he designed and currently oversees nine successfully operating makerspaces across Wyoming, making him a state-recognized authority on makerspace development and programming. For his work, Kerr was a recipient of the 2018 Laramie Young Professionals 20 under 40 award, the University of Wyoming's 2020 Employee of the Quarter award, and the 2021 Employee of the Year award. His academic interests include 3D printing, digitization, and developing creative, gamified, out-of-the-box nerdy ways to engage communities and teach complex topics in meaningful ways. With over 11 years of experience as an outreach coordinator and academic educator in Science, Technology, Engineering, Arts, and Math (STEAM), he aims to prove that everyone and anyone—even paleontological fossils like him—can be a maker.

Steven Barrett Ph.D., P.E., received his BS in Electronic Engineering Technology from the University of Nebraska in Omaha in 1979, M.E.E.E. from the University of Idaho in Moscow in 1986, and Ph.D. from The University of Texas in Austin in 1993. He was formally an active duty faculty member at the United States Air Force Academy, Colorado, and is now Vice Provost of Undergraduate Education at the University of Wyoming and Professor of Electrical and Computer Engineering. He is a member of IEEE (senior) and Tau Beta Pi (chief faculty advisor). His research interests include image processing, computer-assisted laser surgery, and embedded controller systems. He is a registered Professional Engineer in Wyoming and Colorado. In 2004, Barrett was named "Wyoming Professor of the Year" by the Carnegie Foundation for the Advancement of Teaching, and in 2008 he was the recipient of the National Society of Professional Engineers (NSPE) Professional Engineers in Higher Education, Engineering Education Excellence Award.

Getting Started

1

Objectives: After reading this chapter, the reader should be able to do the following:

- Successfully download and execute a simple program using the Arduino Development Environment;
- Describe the key features of the Arduino Development Environment;
- List the programming support information available at the Arduino home page;
- Write programs for use on the Arduino UNO R3;
- Describe the Arduino concept of open source hardware;
- Diagram the layout of the Arduino UNO R3 processor board;
- Name and describe the different features aboard the Arduino UNO R3 processor board;
- Discuss the features and functions of the Microchip ATmega328;
- Diagram the layout of the Arduino Mega 2560 processor board;
- Name and describe the different features aboard the Arduino Mega 2560 R3 processor board;
- Discuss the features and functions of the Microchip ATmega2560; and
- Describe how to extend the hardware features of the Arduino processor using Arduino Shields.

1.1   Overview

Welcome to the world of Arduino! The Arduino concept of open source hardware was developed by the visionary Arduino team of Massimo Banzi, David Cuartilles, Tom Igoe, Gianluca Martino, and David Mellis in Ivrea, Italy. The team's goal was to develop a line of

© The Author(s), under exclusive license to Springer Nature Switzerland AG 2022

1

T. Kerr and S. Barrett, *Arduino IV: DIY Robots*, Synthesis Lectures on Digital Circuits & Systems, https://doi.org/10.1007/978-3-031-11209-6_1

easy–to–use microcontroller hardware and software such that processing power would be readily available to everyone.[1]

This chapter provides a tutorial on the Arduino programming environment and the Arduino UNO R3 and the ATmega 2560 hardware platforms. As you begin your Arduino adventure, you will find yourself coming back to this chapter frequently for reference. If you are a seasoned Arduino user you may want to quickly review the chapter and move on to 3D printing concepts provided in Chap. 2.

Early in the chapter we concentrate on the Arduino programming environment. To the novice, programming a microcontroller may appear mysterious, complicated, overwhelming, and difficult. When faced with a new task, one often does not know where to start. Our goal is to provide a tutorial on how to begin programming. We will use a top–down design approach. We begin with a "big picture" of programming an Arduino and then discuss the Arduino Development Environment and how it may be used to quickly develop a program (sketch) for the Arduino UNO R3. Throughout the chapter, we provide examples and also provide pointers to a number of excellent references. To get the most out of this chapter, it is highly recommended to work through each example.

After getting acquainted with the Arduino IDE, we spend the second portion of the chapter exploring the hardware features and capability of the Arduino UNO R3 and the Mega 2560 R3.

It is important to master both the software and hardware concepts presented in this chapter.

1.2   The Big Picture

We begin with the big picture of how to program the Arduino development boards as shown in Fig. 1.1. This will help provide an overview of how chapter concepts fit together.

Most microcontrollers are programmed with some variant of the C programming language. The C programming language provides a nice balance between the programmer's control of the microcontroller hardware and time efficiency in program writing. As an alternative, the Arduino Development Environment (ADE) provides a user–friendly interface to quickly develop a program or sketch, transform the sketch to machine code, and then load the machine code into the Arduino processor in several simple steps. We use the ADE throughout the book.

The first version of the Arduino Development Environment was released in August 2005. It was developed at the Interaction Design Institute in Ivrea, Italy to allow students the ability to quickly put processing power to use in a wide variety of projects. Since that time, updated versions incorporating new features, have been released on a regular basis (www.arduino. cc).

---

[1] This chapter is a compilation of Chaps. 1 and 2 from "Arduino I: Getting Started." It is included here with permission to provide a complete, standalone text.

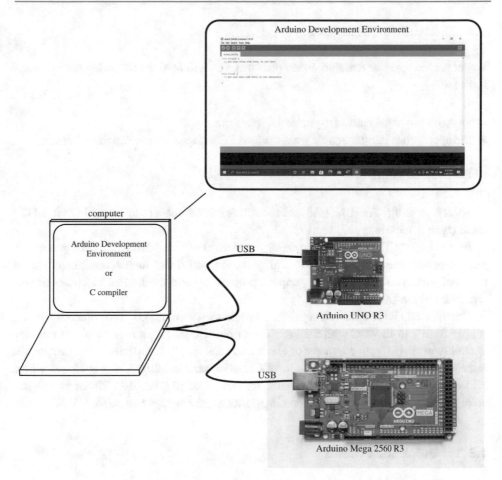

Fig. 1.1 Programming the Arduino processor board. Arduino illustrations used with permission of the Arduino Team (CC BY–NC–SA) (www.arduino.cc)

At its most fundamental level, the Arduino Development Environment is a user–friendly interface to allow one to quickly write, load, and execute code on a microcontroller. A bare-bones program need only consist of a setup() and loop() function. The Arduino Development Environment adds the other required pieces such as header files and the main program construct. The ADE is written in Java and has its origins in the Processor programming language and the Wiring Project (www.arduino.cc).

534069_1_En_1_Chapter-print ☑ TYPESET ☐ DISK ☐ LE ☑ CP Disp.:2/9/2022 Pages: 230 Layout: German_T5

1.3   Arduino Quickstart

To get started using an Arduino–based platform, you will need the following hardware and software.

- an Arduino–based hardware processing platform,
- the appropriate interface cable from the host PC or laptop to the Arduino platform,
- an Arduino compatible power supply, and
- the Arduino software.

Interface cable. The UNO R3 and the ATmega2560 connect to a host PC via a USB cable (type A male to type B female).

Power supply. The Arduino processing board may be powered from the USB port during project development. However, it is highly recommended that an external power supply be employed. This will allow developing projects beyond the limited electrical current capability of the USB port.

For the UNO R3 and the ATmega2560, Arduino www.arduino.cc recommends a power supply from 7 to 12 VDC with a 2.1 mm center positive plug. A power supply of this type is readily available from a number of electronic parts supply companies. For example, the Jameco #133891 power supply is a 9 VDC model rated at 300 mA and equipped with a 2.1 mm center positive plug. It is available for under US$10. The UNO R3 has an onboard voltage regulator to maintain the incoming power supply voltage at a stable 5 VDC.

1.3.1   Quick Start Guide

The Arduino Development Environment may be downloaded from the Arduino website's front page at www.arduino.cc [1]. Versions are available for Windows, Mac OS X, and Linux. Provided below is a quick start step–by–step approach to blink an onboard LED.

- Download the Arduino Development Environment from www.arduino.cc [1].
- Connect the Arduino UNO R3 processing board to the host computer via a USB cable (A male to B male).
- Start the Arduino Development Environment.
- Under the Tools tab select the evaluation **Board** you are using and the **Port** that it is connected to.
- Type the following program.

```
//***************************************************************

#define LED_PIN 13

void setup()
{
pinMode(LED_PIN, OUTPUT);
}

void loop()
{
digitalWrite(LED_PIN, HIGH);
delay(500);                        //delay specified in ms
digitalWrite(LED_PIN, LOW);
delay(500);
}

//***************************************************************
```

- Upload and execute the program by asserting the "Upload" (right arrow) button.
- The onboard LED should blink at one second intervals.

With the Arduino ADE downloaded and exercised, let's take a closer look at its features.

1.3.2   Arduino Development Environment Overview

The Arduino Development Environment is illustrated in Fig. 1.2. The ADE contains a text editor, a message area for displaying status, a text console, a tool bar of common functions, and an extensive menuing system. The ADE also provides a user–friendly interface to the Arduino processor board which allows for a quick upload of code. This is possible because the Arduino processing boards are equipped with a bootloader program.

A close up of the Arduino toolbar is provided in Fig. 1.3. The toolbar provides single button access to the more commonly used menu features. Most of the features are self–explanatory. As described in the previous section, the "Upload" button compiles your code and uploads it to the Arduino processing board. The "Serial Monitor" button opens the serial monitor feature. The serial monitor feature allows text data to be sent to and received from the Arduino processing board.

1.3.3   Sketchbook Concept

In keeping with a hardware and software platform for students of the arts, the Arduino environment employs the concept of a sketchbook. An artist maintains their works in progress

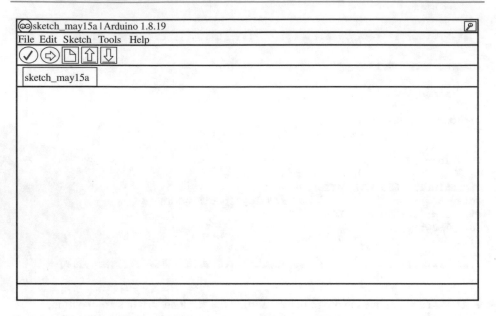

Fig. 1.2 Arduino Development Environment (www.arduino.cc)

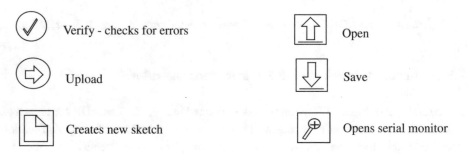

Fig. 1.3 Arduino Development Environment buttons

in a sketchbook. Similarly, we maintain our programs within a sketchbook in the Arduino environment. Furthermore, we refer to individual programs as sketches. An individual sketch within the sketchbook may be accessed via the Sketchbook entry under the file tab.

1.3.4   Arduino Software, Libraries, and Language References

The Arduino Development Environment has a number of built–in features. Some of the features may be directly accessed via the Arduino Development Environment drop down toolbar illustrated in Fig. 1.2. Provided in Fig. 1.4 is a handy reference to show the available features. The toolbar provides a wide variety of features to compose, compile, load and execute a sketch.

534069_1_En_1_Chapter-print ☑ TYPESET ☐ DISK ☐ LE ☑ CP Disp.:2/9/2022 Pages: 230 Layout: German_T5

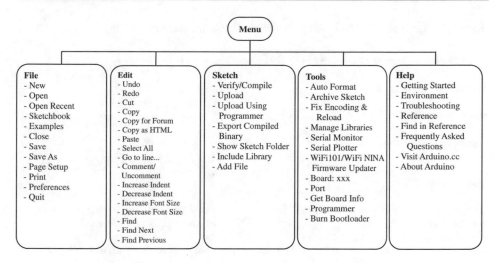

Fig. 1.4 Arduino Development Environment menu (www.arduino.cc)

1.3.5 Writing an Arduino Sketch

The basic format of the Arduino sketch consists of a "setup" and a "loop" function. The setup function is executed once at the beginning of the program. It is used to configure pins, declare variables and constants, etc. The loop function will execute sequentially step–by––step. When the end of the loop function is reached it will automatically return to the first step of the loop function and execute again. This goes on continuously until the program is stopped.

```
//***********************************************************

void setup()
  {
  //place setup code here
  }

void loop()
  {
  //main code steps are provided here
  :
  :

  }

//***********************************************************
```

Example: Let's revisit the sketch provided earlier in the chapter.

```
//************************************************************

#define LED_PIN 13                          //name pin 13 LED_PIN

void setup()
{
pinMode(LED_PIN, OUTPUT);                    //set pin to output
}

void loop()
{
digitalWrite(LED_PIN, HIGH);                 //write pin to logic high
delay(500);                                  //delay specified in ms
digitalWrite(LED_PIN, LOW);                  //write to logic low
delay(500);                                  //delay specified in ms
}

//************************************************************
```

In the first line the #define statement links the designator "LED_PIN" to pin 13 on the Arduino processor board. In the setup function, LED_PIN is designated as an output pin. Recall the setup function is only executed once. The program then enters the loop function that is executed sequentially step–by–step and continuously repeated. In this example, the LED_PIN is first set to logic high to illuminate the LED onboard the Arduino processing board. A 500 ms delay then occurs. The LED_PIN is then set low. A 500 ms delay then occurs. The sequence then repeats.

Even the most complicated sketches follow the basic format of the setup function followed by the loop function. To aid in the development of more complicated sketches, the Arduino Development Environment has many built–in features that may be divided into the areas of structure, variables and functions. The structure and variable features follow rules similar to the C programming language which is discussed in the text "Arduino II: Systems."[2] The built–in functions consists of a set of pre–defined activities useful to the programmer. These built–in functions are summarized in Fig. 1.5.

There are many program examples available to allow the user to quickly construct a sketch. These programs are summarized in Fig. 1.6. Complete documentation for these programs is available at the Arduino homepage (www.arduino.cc). This documentation is easily accessible via the Help tab on the Arduino Development Environment toolbar. This documentation will not be repeated here. Instead, we refer to these features at appropriate places throughout the remainder of the book. With the Arduino open source concept, users

---

[2] S. F. Barrett, "Arduino II: Systems," Morgan and Claypool Publishers, 2020.

534069_1_En_1_Chapter-print ☑ TYPESET ☐ DISK ☐ LE ☑ CP Disp.:2/9/2022 Pages: 230 Layout: German_T5

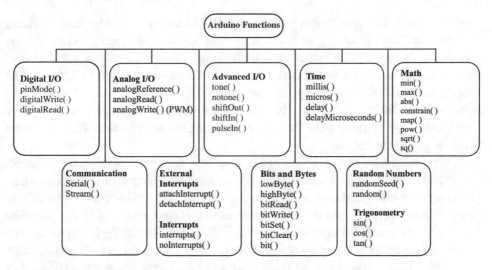

Fig. 1.5 Arduino Development Environment functions (www.arduino.cc)

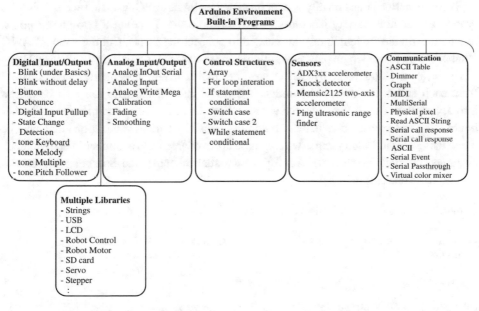

Fig. 1.6 Arduino Development Environment built–in features (www.arduino.cc)

throughout the world are constantly adding new built–in features. As new features are added, they will be released in future Arduino Development Environment versions. As an Arduino user, you too may add to this collection of useful tools. We continue with another example. **Example:** In this example we connect an external LED to Arduino UNO R3 pin 12. The onboard LED will blink alternately with the external LED. The external LED is connected to the Arduino UNO R3 as shown in Fig. 1.7. The LED has a positive (anode) and negative (cathode) lead. As you look down from above, a round LED has a flat side. The lead closest to the flat side is the cathode.

In the bottom right portion of the figure, a cutaway is provided of a prototype board. Prototype boards provide a convenient method of interconnecting electronic components. The boards are configured to accept dual inline package (DIP) integrated circuits (ICs or chips). The ICs are placed over the center channels. Other components may also be placed on the boards. The boards are covered with multiple connection points. Typically, the board's connection points are arranged in columns and five connection point rows. The connection points in a column are connected at the board's base by a conductor. Also, connection points in a board row are connected.

The connection points readily accept insulated, solid 22 AWG insulated wire. This wire type is available in variety of colors (www.jameco.com). To connect two circuit points together, estimate the length of wire needed to connect the points. Cut the solid 22 AWG insulated wire to that length plus an additional one–half inch (approximately 13 mm). Using a wire stripper (e.g. Jameco #159291), strip off approximately one–quarter inch (7.5 mm) from each end of the wire. The conductor exposed ends of the wire can then be placed into appropriate circuit locations for a connection. Once a wire end is placed in a bread-board hole, the exposed conductor should not be visible. Circuit connections may also be made using prefabricated jumper wires. These are available from a number of sources such as Jameco (www.jameco.com), Adafruit (www.adafruit.com), and SparkFun Electronics (www.sparkfun.com) [4].

```
//************************************************************

#define int_LED 13                        //name pin 13 int_LED
#define ext_LED 12                        //name pin 12 ext_LED

void setup()
{
pinMode(int_LED, OUTPUT);                 //set pin to output
pinMode(ext_LED, OUTPUT);                 //set pin to output
}

void loop()
{
digitalWrite(int_LED, HIGH);              //set pin logic high
digitalWrite(ext_LED, LOW);               //set pin logic low
delay(500);                               //delay specified in ms
digitalWrite(int_LED, LOW);               //set pin logic low
```

```
digitalWrite(ext_LED, HIGH);            //set pin logic high
delay(500);
}


//*********************************************************
```

Example: In this example we connect an external LED to Arduino UNO R3 pin 12 and an external switch attached to pin 11. The onboard LED will blink alternately with the external LED when the switch is depressed. The external LED and switch is connected to the Arduino UNO R3 as shown in Fig. 1.8.

```
//*********************************************************


#define int_LED 13                      //name pin 13 int_LED
#define ext_LED 12                      //name pin 12 ext_LED
#define ext_sw  11                      //name pin 11 ext_sw

int switch_value;                       //integer variable to
                                        //store switch status

void setup()
{
pinMode(int_LED, OUTPUT);               //set pin to output
pinMode(ext_LED, OUTPUT);               //set pin to output
pinMode(ext_sw,  INPUT);                //set pin to input
}

void loop()
{
switch_value = digitalRead(ext_sw);     //read switch status
if(switch_value == LOW)                 //if switch at logic low,
  {                                     //do steps with braces
  digitalWrite(int_LED, HIGH);          //set pin logic high
  digitalWrite(ext_LED, LOW);           //set pin logic low
  delay(50);                            //delay 50 ms
  digitalWrite(int_LED, LOW);           //set pin logic low
  digitalWrite(ext_LED, HIGH);          //set pin logic high
  delay(50);                            //delay 50ms
  }
else                                    //if switch at logic high,
  {                                     //do steps between braces
  digitalWrite(int_LED, LOW);           //set pins low
  digitalWrite(ext_LED, LOW);           //set pins low
  }
}


//*********************************************************
```

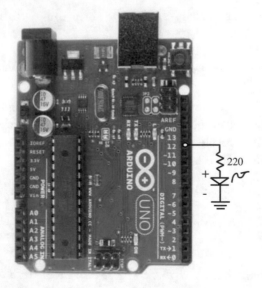

a) schematic

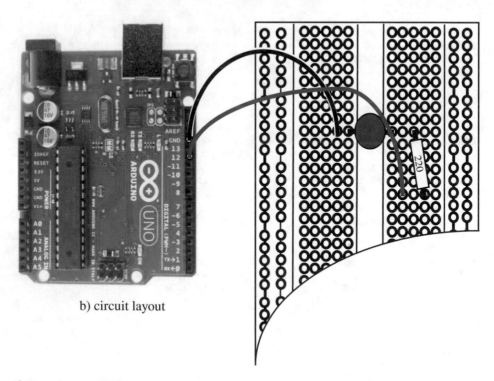

b) circuit layout

Fig. 1.7 Arduino UNO R3 with an external LED. UNO R3 illustration used with permission of the Arduino Team (CC BY–NC–SA) (www.arduino.cc)

534069_1_En_1_Chapter-print ☑ TYPESET ☐ DISK ☐ LE ☑ CP Disp.:**2/9/2022** Pages: **230** Layout: **German_T5**

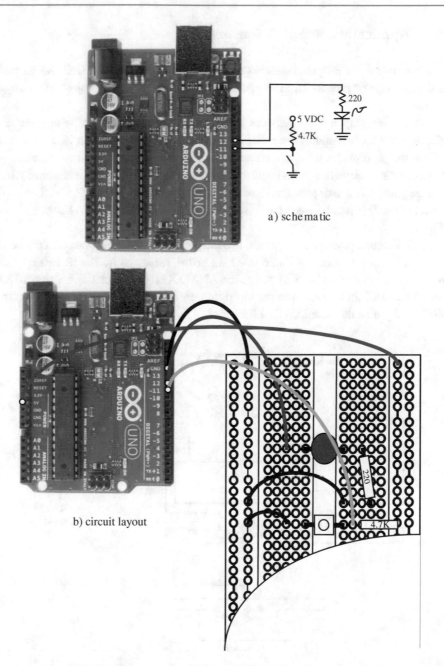

a) schematic

b) circuit layout

Fig. 1.8 Arduino UNO R3 with an external LED. UNO R3 illustration used with permission of the Arduino Team (CC BY–NC–SA) (www.arduino.cc)

1.4   Application: Robot IR Sensor

In this example we investigate a sketches's interaction with the Arduino UNO R3 processing board and external sensors and indicators. We will use the robot project as an ongoing example.

Later in the chapter we equip the Dagu Rover 5 robot platform with three Sharp GP2Y0A41SK0F (we abbreviate as GP2Y) infrared (IR) sensors. The IR sensor, with a 4–30 cm range, provides a voltage output that is inversely proportional to the sensor distance from the maze wall. It is designed to illuminate the LED if the robot is within 10 cm of the maze wall. The sensor provides an output voltage of approximately 2.5 VDC at the 10–cm range. The interface between the IR sensor and the Arduino UNO R3 board is provided in Fig. 1.9.

The IR sensor's power (red wire) and ground (black wire) connections are connected to the 5V and Gnd pins on the Arduino UNO R3 board, respectively. The IR sensor's output connection (yellow wire) is connected to the ANALOG IN 5 pin on the Arduino UNO R3 board. The LED circuit shown in the top right corner of the diagram is connected to the DIGITAL 0 pin on the Arduino UNO R3 board.

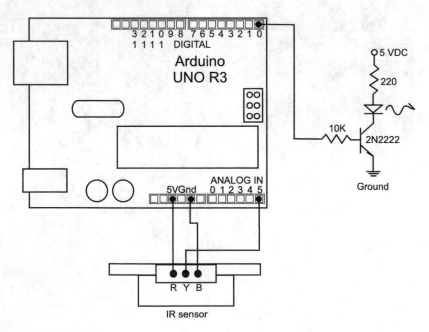

Fig. 1.9 IR sensor interface

```
//*******************************************************************
#define LED_PIN    0               //digital pin - LED connection
#define IR_sensor_pin 5            //analog pin - IR sensor

int IR_sensor_value;              //declare variable for IR sensor value

void setup()
  {
    pinMode(LED_PIN, OUTPUT);      //configure pin 0 for digital output
  }

void loop()
  {
                                   //read analog output from IR sensor
    IR_sensor_value = analogRead(IR_sensor_pin);

    if(IR_sensor_value > 512)      //0 to 1023 maps to 0 to 5 VDC
      {
      digitalWrite(LED_PIN, HIGH); //turn LED on
      }
    else
      {
      digitalWrite(LED_PIN, LOW);  //turn LED off
      }
  }
//*******************************************************************
```

The sketch begins by providing names for the two Arduino UNO R3 board pins that will be used in the sketch. This is not required but it makes the code easier to read. We define the pin for the LED as "LED_PIN." Any descriptive name may be used here. Whenever the name is used within the sketch, the number "0" will be substituted for the name by the compiler.

After providing the names for pins, the next step is to declare any variables required by the sketch. In this example, the output from the IR sensor will be converted from an analog to a digital value using the built–in Arduino "analogRead" function. A detailed description of the function may be accessed via the Help menu. It is essential to carefully review the support documentation for a built–in Arduino function the first time it is used. The documentation provides details on variables required by the function, variables returned by the function, and an explanation on function operation.

The "analogRead" function requires the pin for analog conversion variable passed to it and returns the analog signal read as an integer value (int) from 0 to 1023. So, for this example, we need to declare an integer value to receive the returned value. We have called this integer variable "IR_sensor_value."

Following the declaration of required variables are the two required functions for an Arduino UNO R3 program: setup and loop. The setup function calls an Arduino built–in function, pinMode, to set the "LED_PIN" as an output pin. The loop function calls several functions to read the current analog value on pin 5 (the IR sensor output) and then determine if the reading is above 512 (2.5 VDC). If the reading is above 2.5 VDC, the LED on DIGITAL pin 0 is illuminated, else it is turned off.

After completing writing the sketch with the Arduino Development Environment, it must be compiled and then uploaded to the Arduino UNO R3 board. These two steps are accomplished using the "Sketch – Verify/Compile" and the "File – Upload to I/O Board" pull down menu selections.

Exercise: Develop a range versus voltage plot for the IR sensor for ranges from 0 to 40 cm.

1.5   Application: External Interrupts

The interrupt system onboard a microcontroller allows it to respond to higher priority events. Appropriate responses to these events may be planned, but we do not know when these events will occur. When an interrupt event occurs, the microcontroller will normally complete the instruction it is currently executing and then transition program control to interrupt event specific tasks. These tasks, which resolve the interrupt event, are organized into a function called an interrupt service routine (ISR). Each interrupt will normally have its own interrupt specific ISR. Once the ISR is complete, the microcontroller will resume processing where it left off before the interrupt event occurred (Fig. 1.10).

The Arduino Development Environment has four built–in functions to support external the INT0 and INT1 external interrupts (www.arduino.cc).

These are the four functions:

Fig. 1.10 Microcontroller
Interrupt Response

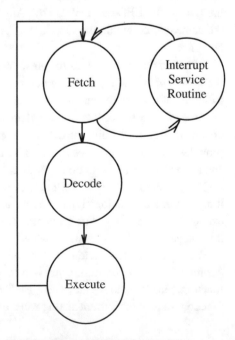

- **interrupts()**. This function enables interrupts.
- **noInterrupts()**. This function disables interrupts.
- **attachInterrupt(interrupt, function, mode)**. This function links the interrupt to the appropriate interrupt service routine.
- **detachInterrupt(interrupt)**. This function turns off the specified interrupt.

The Arduino UNO R3 processing board is equipped with two external interrupts: INT0 on pin 2 and INT1 on pin 3.

The **attachInterrupt(interrupt, function, mode)** function is used to link the hardware pin to the appropriate interrupt service pin. The three arguments of the function are configured as follows:

- **interrupt.** Interrupt specifies the INT interrupt number: either 0 or 1.
- **function.** Function specifies the name of the interrupt service routine.
- **mode.** Mode specifies what activity on the interrupt pin will initiate the interrupt: **LOW level on pin, CHANGE in pin level, RISING edge, or FALLING edge.**

Provided below is a template to configure an interrupt.

```
//*************************************************************

void setup()
{
attachInterrupt(0, int0_ISR, FALLING);
}

void loop()
{

//wait for interrupts

}

//*************************************************************
//int0_ISR: interrupt service routine for INT0
//*************************************************************

void int0_ISR(void)
{

//Insert interrupt specific actions here.

}
//*************************************************************
```

As an example, we connect an external tact switch to INT0 (pin 2) and measure the elapsed time between two switch presses. The switch configuration provided earlier in Fig. 1.8 may be used.

534069_1_En_1_Chapter-print ☑ TYPESET ☐ DISK ☐ LE ☑ CP Disp.:2/9/2022 Pages: 230 Layout: German_T5

```
//******************************************************************
//Program measures the elapsed time in ms between two switch
//closures.  A tact switch with a series 4.7K resistor is attached
//to INT0 (pin 2) of the UNO R3.
//******************************************************************

unsigned long first, second, elapsed_time;
unsigned int first_time_hack = 1;

void setup()
{
Serial.begin(9600);
pinMode(2, INPUT);
attachInterrupt(0, int0_ISR, FALLING);
}

void loop()
{

//wait for interrupts

}

//******************************************************************
//int0_ISR: interrupt service routine for INT0
//******************************************************************

void int0_ISR(void)
{
if(first_time_hack ==1)
   {
   first = millis();
   first_time_hack = 0;
   delay(5);
   }
else
   {
   second = millis();
   first_time_hack = 1;
   elapsed_time = second  first;
   Serial.print(elapsed_time);
   Serial.println(" ms");
   Serial.println();
   delay(5);
   }
}

//******************************************************************
```

1.6   Arduino UNO R3 Processing Board

The Arduino UNO R3 processing board is illustrated in Fig. 1.11. Working clockwise from the left, the board is equipped with a USB connector to allow programming the processor from a host personal computer (PC) or laptop. The board may also be programmed using In System Programming (ISP) techniques. A 6–pin ISP programming connector is on the opposite side of the board from the USB connector.

The board is equipped with a USB–to–serial converter to allow compatibility between the host PC and the serial communications systems aboard the Microchip ATmega328 processor. The UNO R3 is also equipped with several small surface mount light emitting diodes (LEDs) to indicate serial transmission (TX) and reception (RX) and an extra LED for project use. The header strip at the top of the board provides access for an analog reference signal, pulse width modulation (PWM) signals, digital input/output (I/O), and serial communications. The header strip at the bottom of the board provides analog inputs for the analog–to–digital (ADC) system and power supply terminals. Finally, the external power supply connector is provided at the bottom left corner of the board. The top and bottom header strips conveniently mate with an Arduino shield to extend the features of the Arduino host processor.

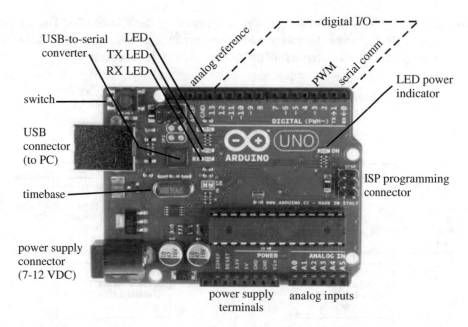

Fig. 1.11 Arduino UNO R3 layout. Figure adapted and used with permission of Arduino Team (CC BY–NC–SA)(www.arduino.cc)

534069_1_En_1_Chapter-print ☑ TYPESET ☐ DISK ☐ LE ☑ CP Disp.:2/9/2022 Pages: 230 Layout: German_T5

1.7   Advanced: Arduino UNO R3 Host Processor–The ATmega328

The host processor for the Arduino UNO R3 is the Microchip Atmega328. The "328" is a 28 pin, 8–bit microcontroller. The architecture is based on the Reduced Instruction Set Computer (RISC) concept which allows the processor to complete 20 million instructions per second (MIPS) when operating at 20 MHz. The "328" is equipped with a wide variety of features as shown in Fig. 1.12. The features may be conveniently categorized into the following systems:

- Memory system,
- Port system,
- Timer system,
- Analog–to–digital converter (ADC),
- Interrupt system, and
- Serial communications.

1.7.1   Arduino UNO R3/ATmega328 Hardware Features

The Arduino UNO R3's processing power is provided by the ATmega328. The pin out diagram and block diagram for this processor are provided in Figs. 1.13 and 1.14. In this section, we provide a brief overview of the systems aboard the processor.

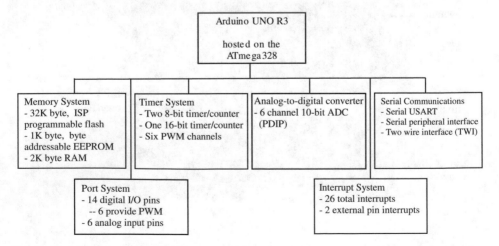

Fig. 1.12 Arduino UNO R3 systems

534069_1_En_1_Chapter-print ☑ TYPESET ☐ DISK ☐ LE ☑ CP Disp.:2/9/2022 Pages: 230 Layout: German_T5

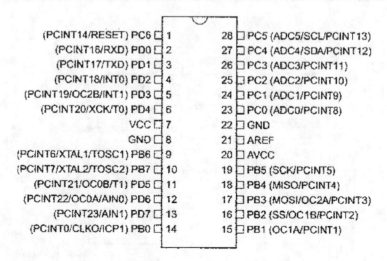

Fig. 1.13 ATmega328 pin out. Figure used with permission of Microchip, Incorporated (www. microchip.com) [2]

1.7.2   ATmega328 Memory

The ATmega328 is equipped with three main memory sections: flash electrically erasable programmable read only memory (EEPROM), static random access memory (SRAM), and byte–addressable EEPROM. We discuss each memory component in turn.

1.7.2.1 ATmega328 In–System Programmable Flash EEPROM
Bulk programmable flash EEPROM is used to store programs. It can be erased and programmed as a single unit. Also, should a program require a large table of constants, it may be included as a global variable within a program and programmed into flash EEPROM with the rest of the program. Flash EEPROM is nonvolatile meaning memory contents are retained even when microcontroller power is lost. The ATmega328 is equipped with 32 K bytes of onboard reprogrammable flash memory. This memory component is organized into 16 K locations with 16 bits at each location.

1.7.2.2 ATmega328 Byte–Addressable EEPROM
Byte–addressable EEPROM memory is used to permanently store and recall variables during program execution. It too is nonvolatile. It is especially useful for logging system malfunctions and fault data during program execution. It is also useful for storing data that must be retained during a power failure but might need to be changed periodically. Examples where this type of memory is used are found in applications to store system parameters,

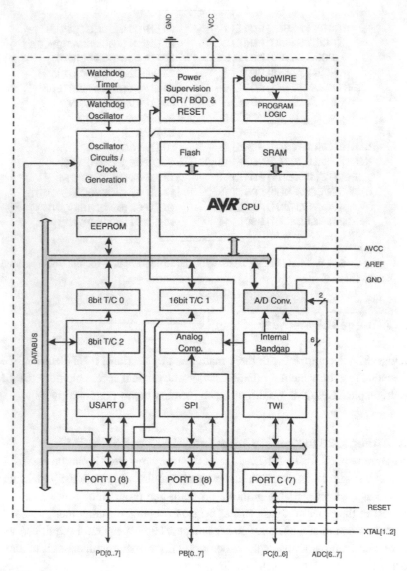

Fig. 1.14 ATmega328 block diagram. Figure used with permission of Microchip, Incorporated (www. microchip.com) [2]

534069_1_En_1_Chapter-print ☑ TYPESET ☐ DISK ☐ LE ☑ CP Disp.:**2/9/2022** Pages: **230** Layout: **German_T5**

electronic lock combinations, and automatic garage door electronic unlock sequences. The ATmega328 is equipped with 1024 bytes of EEPROM.

1.7.2.3 ATmega328 Static Random Access Memory (SRAM)

Static RAM memory is volatile. That is, if the microcontroller loses power, the contents of SRAM memory are lost. It can be written to and read from during program execution. The ATmega328 is equipped with 2 K bytes of SRAM. A small portion of the SRAM is set aside for the general–purpose registers used by the processor and also for the input/output and peripheral subsystems aboard the microcontroller. A header file provides the link between register names used in a program and their physical description and location in memory. During program execution, RAM is used to store global variables, support dynamic memory allocation of variables, and to provide a location for the stack.

1.7.3   ATmega328 Port System

The Microchip ATmega328 is equipped with four, 8–bit general purpose, digital input/output (I/O) ports designated PORTB (8 bits, PORTB[7:0]), PORTC (7 bits, PORTC[6:0]), and PORTD (8 bits, PORTD[7:0]). As shown in Fig. 1.21, each port has three registers associated with it:

- Data Register PORTx—used to write output data to the port,
- Data Direction Register DDRx—used to set a specific port pin to either output (1) or input (0), and
- Input Pin Address PINx—used to read input data from the port (Fig. 1.15).

Figure 1.21b describes the settings required to configure a specific port pin to either input or output. If selected for input, the pin may be selected for either an input pin or to operate in the high impedance (Hi–Z) mode. If selected for output, the pin may be further configured for either logic low or logic high.

Port pins are usually configured at the beginning of a program for either input or output and their initial values are then set. Usually all eight pins for a given port are configured simultaneously.

1.7.4   ATmega328 Internal Systems

In this section, we provide a brief overview of the internal features of the ATmega328. It should be emphasized that these features are the internal systems contained within the confines of the microcontroller chip. These built–in features allow complex and sophisticated tasks to be accomplished by the microcontroller.

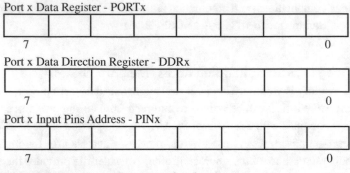

Port x Data Register - PORTx

7   0

Port x Data Direction Register - DDRx

7   0

Port x Input Pins Address - PINx

7   0

a) port associated registers

DDxn	PORTxn	I/O	Comment	Pullup
0	0	input	Tri-state (Hi-Z)	No
0	1	input	source current if externally pulled low	Yes
1	0	output	Output Low (Sink)	No
1	1	output	Output High (Source)	No

x: port designator (B, C, D)
n: pin designator (0 - 7)

b) port pin configuration

Fig. 1.15 ATmega328 port configuration registers

1.7.4.1 ATmega328 Time Base

The microcontroller is a complex synchronous state machine. It responds to program steps in a sequential manner as dictated by a user–written program. The microcontroller sequences through a predictable fetch–decode–execute sequence. Each unique assembly language program instruction issues a series of signals to control the microcontroller hardware to accomplish instruction related operations.

The speed at which a microcontroller sequences through these actions is controlled by a precise time base called the clock. The clock source is routed throughout the microcontroller to provide a time base for all peripheral subsystems. The ATmega328 may be clocked internally using a user–selectable resistor capacitor (RC) time base or it may be clocked externally. The RC internal time base is selected using programmable fuse bits. You may choose from several different internal fixed clock operating frequencies.

To provide for a wider range of frequency selections an external time source may be used. The external time sources, in order of increasing accuracy and stability, are an external RC network, a ceramic resonator, or a crystal oscillator. The system designer chooses the time base frequency and clock source device appropriate for the application at hand. Generally speaking, if the microcontroller will be interfaced to external peripheral devices either a ceramic resonator or a crystal oscillator should be used as a time base.

1.7.4.2 ATmega328 Timing Subsystem
The ATmega328 is equipped with a complement of timers which allows the user to generate a precision output signal, measure the characteristics (period, duty cycle, frequency) of an incoming digital signal, or count external events. Specifically, the ATmega328 is equipped with two 8–bit timer/counters and one 16–bit counter.

1.7.4.3 Pulse Width Modulation Channels
A pulse width modulated or PWM signal is characterized by a fixed frequency and a varying duty cycle. Duty cycle is the percentage of time a repetitive signal is logic high during the signal period. It may be formally expressed as:

$$duty\ cycle[\%] = (on\ time/period) \times (100\%)$$

The ATmega328 is equipped with four pulse width modulation (PWM) channels. The PWM channels coupled with the flexibility of dividing the time base down to different PWM subsystem clock source frequencies allows the user to generate a wide variety of PWM signals: from relatively high frequency low duty cycle signals to relatively low frequency high duty cycle signals.

PWM signals are used in a wide variety of applications including controlling the position of a servo motor and controlling the speed of a DC motor.

1.7.4.4 ATmega328 Serial Communications
The ATmega328 is equipped with a variety of different serial communication subsystems including the Universal Synchronous and Asynchronous Serial Receiver and Transmitter (USART), the serial peripheral interface (SPI), and the Two–wire Serial Interface. What these systems have in common is the serial transmission of data. In a serial communications transmission, serial data is sent a single bit at a time from transmitter to receiver.

ATmega328 Serial USART The serial USART may be used for full duplex (two way) communication between a receiver and transmitter. This is accomplished by equipping the ATmega328 with independent hardware for the transmitter and receiver. The USART is typically used for asynchronous communication. That is, there is not a common clock between the transmitter and receiver to keep them synchronized with one another. To maintain synchronization between the transmitter and receiver, framing start and stop bits are used at the beginning and end of each data byte in a transmission sequence.

The ATmega328 USART is quite flexible. It has the capability to be set to different data transmission rates known as the Baud (bits per second) rate. The USART may also be set for data bit widths of 5–9 bits with one or two stop bits. Furthermore, the ATmega328 is equipped with a hardware generated parity bit (even or odd) and parity check hardware at the receiver. A single parity bit allows for the detection of a single bit error within a byte of data. The USART may also be configured to operate in a synchronous mode.

ATmega328 Serial Peripheral Interface–SPI The ATmega328 Serial Peripheral Interface (SPI) can also be used for two–way serial communication between a transmitter and a receiver. In the SPI system, the transmitter and receiver share a common clock source. This requires an additional clock line between the transmitter and receiver but allows for higher data transmission rates as compared to the USART.

The SPI may be viewed as a synchronous 16–bit shift register with an 8–bit half residing in the transmitter and the other 8–bit half residing in the receiver. The transmitter is designated the master since it is providing the synchronizing clock source between the transmitter and the receiver. The receiver is designated as the slave.

ATmega328 Two–wire Serial Interface–TWI The TWI subsystem allows the system designer to network related devices (microcontrollers, transducers, displays, memory storage, etc.) together into a system using a two–wire interconnecting scheme. The TWI allows a maximum of 128 devices to be interconnected. Each device has its own unique address and may both transmit and receive over the two–wire bus at frequencies up to 400 kHz. This allows the device to freely exchange information with other devices in the network within a small area.

1.7.4.5 ATmega328 Analog to Digital Converter–ADC
The ATmega328 is equipped with an eight–channel analog to digital converter (ADC) subsystem. The ADC converts an analog signal from the outside world into a binary representation suitable for use by the microcontroller. The ATmega328 ADC has 10–bit resolution. This means that an analog voltage between 0 and 5 V will be encoded into one of 1024 binary representations between $(000)_{16}$ and $(3FF)_{16}$. This provides the ATmega328 with a voltage resolution of approximately 4.88 mV.

1.7.4.6 ATmega328 Interrupts
The normal execution of a program follows a designated sequence of instructions. However, sometimes this normal sequence of events must be interrupted to respond to high priority faults and status both inside and outside the microcontroller. When these higher priority events occur, the microcontroller suspends normal operation and executes event specific actions contained within an interrupt service routine (ISR). Once the higher priority event has been serviced by the ISR, the microcontroller returns and continues processing the normal program.

534069_1_En_1_Chapter-print ☑ TYPESET ☐ DISK ☐ LE ☑ CP Disp.:2/9/2022 Pages: 230 Layout: German_T5

The ATmega328 is equipped with a complement of 26 interrupt sources. Two of the interrupts are provided for external interrupt sources while the remaining interrupts support the efficient operation of peripheral subsystems aboard the microcontroller.

1.8   Arduino UNO R3 Open Source Schematic

The entire line of Arduino products is based on the visionary concept of open source hardware and software. That is, hardware and software developments are openly shared among users to stimulate new ideas and advance the Arduino concept. In keeping with the Arduino concept, the Arduino team openly shares the schematic of the Arduino UNO R3 processing board. Reference Fig. 1.16.

1.9   Arduino Mega 2560 R3 Processing Board

Throughout the book we concentrate on the UNO R3 platform. However, several projects will require additional features (e.g. additional RAM, more interrupts, etc.) beyond the UNO R3. For these projects we employ the Arduino Mega 2560 R3. Concepts learned for the UNO R3 directly port over to the Mega 2560 R3.

The Arduino Mega 2560 REV3 (R3) processing board is illustrated in Fig. 1.17. Working clockwise from the left, the board is equipped with a USB connector to allow programming the processor from a host PC. The board may also be programmed using In System Programming (ISP) techniques. A 6–pin ISP programming connector is on the opposite side of the board from the USB connector.

The board is equipped with a USB–to–serial converter to allow compatibility between the host PC and the serial communications systems aboard the ATmega2560 processor. The Mega 2560 R3 is also equipped with several small surface mount LEDs to indicate serial transmission (TX) and reception (RX) and an extra LED for project use. The header strip at the top of the board provides access to pulse width modulation (PWM) signals and serial communications. The header strip at the right side of the board provides access to multiple digital input/output pins. The bottom of the board provides analog inputs for the analog––to–digital (ADC) system and power supply terminals. Finally, the external power supply connector is provided at the bottom left corner of the board. The header strips conveniently mate with an Arduino shield (to be discussed shortly) to extend the features of the host processor.

534069_1_En_1_Chapter-print ☑ TYPESET ☐ DISK ☐ LE ☑ CP Disp.:2/9/2022 Pages: 230 Layout: German_T5

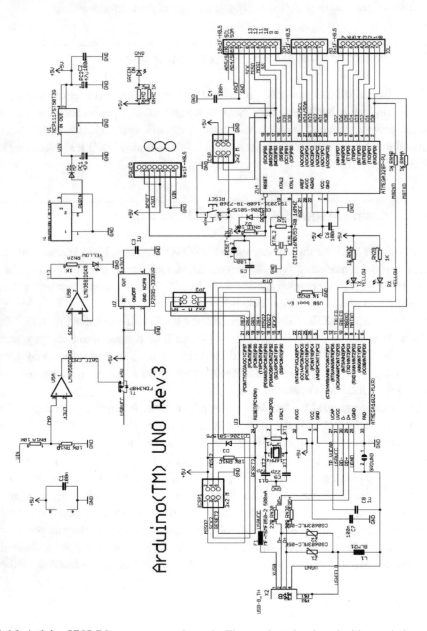

Fig. 1.16 Arduino UNO R3 open source schematic. Figure adapted and used with permission of the Arduino Team (CC BY–NC–SA) (www.arduino.cc)

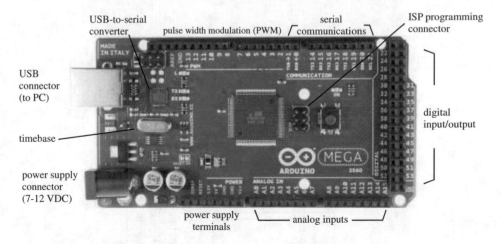

Fig. 1.17 Arduino Mega2560 layout. Figure adapted and used with permission of Arduino Team (CC BY–NC–SA) (www.arduino.cc)

1.10 Advanced: Arduino Mega 2560 Host Processor–The ATmega2560

The host processor for the Arduino Mega 2560 is the Microchip Atmega2560. The "2560" is a 100 pin, surface mount 8–bit microcontroller. The architecture is based on the Reduced Instruction Set Computer (RISC) concept which allows the processor to complete 16 million instructions per second (MIPS) when operating at 16 MHz. The "2560" is equipped with a wide variety of features as shown in Fig. 1.18. The features may be conveniently categorized into the following systems:

- Memory system,
- Port system,
- Timer system,
- Analog–to–digital converter (ADC),
- Interrupt system,
- and serial communications.

1.10.1 Arduino Mega 2560 /ATmega2560 Hardware Features

The Arduino Mega 2560's processing power is provided by the ATmega2560. The pin out diagram and block diagram for this processor are provided in Figs. 1.19 and 1.20. In this section, we provide a brief overview of the systems onboard the processor.

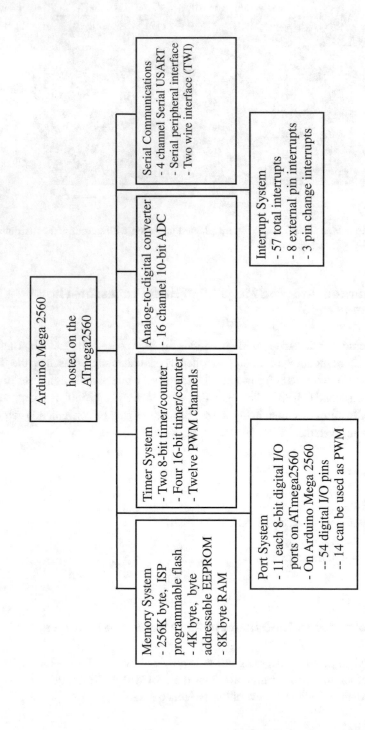

Fig. 1.18 Arduino Mega 2560 systems

534069_1_En_1_Chapter-print ☑ TYPESET ☐ DISK ☐ LE ☑ CP Disp.:**2/9/2022** Pages: **230** Layout: **German_T5**

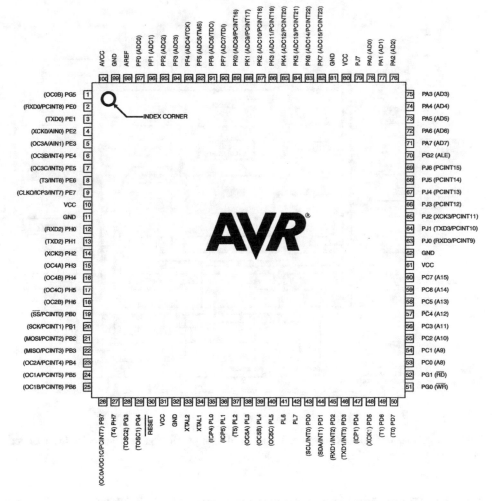

Fig. 1.19 ATmega2560 pin out. Figure used with permission of Microchip, Incorporated (www.microchip.com) [2]

1.10.2 ATmega2560 Memory

The ATmega2560 is equipped with three main memory sections: flash electrically erasable programmable read only memory (EEPROM), static random access memory (SRAM), and byte–addressable EEPROM for data storage.

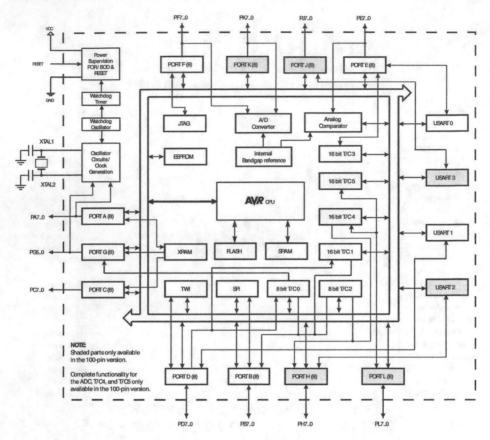

Fig. 1.20 ATmega2560 block diagram. Figure used with permission of Microchip, Incorporated (www.microchip.com) [2]

1.10.2.1 ATmega2560 In–System Programmable Flash EEPROM

Bulk programmable flash EEPROM is used to store programs. It can be erased and programmed as a single unit. Also, should a program require a large table of constants, it may be included as a global variable within a program and programmed into flash EEPROM with the rest of the program. Flash EEPROM is nonvolatile meaning memory contents are retained when microcontroller power is lost. The ATmega2560 is equipped with 256 K bytes of onboard reprogrammable flash memory.

1.10.2.2 ATmega2560 Byte–Addressable EEPROM

Byte–addressable EEPROM memory is used to permanently store and recall variables during program execution. It too is nonvolatile. It is especially useful for logging system malfunctions and fault data during program execution. It is also useful for storing data that must be retained during a power failure but might need to be changed periodically. Examples where this type of memory is used are found in applications to store system parameters, electronic lock combinations, and automatic garage door electronic unlock sequences. The ATmega2560 is equipped with 4096 bytes of byte–addressable EEPROM.

1.10.2.3 ATmega2560 Static Random Access Memory (SRAM)

Static RAM memory is volatile. That is, if the microcontroller loses power, the contents of SRAM memory are lost. It can be written to and read from during program execution. The ATmega2560 is equipped with 8K bytes of SRAM. A small portion of the SRAM is set aside for the general–purpose registers used by the processor and also for the input/output and peripheral subsystems aboard the microcontroller. During program execution, RAM is used to store global variables, support dynamic memory allocation of variables, and to provide a location for the stack.

1.10.3 ATmega2560 Port System

The Microchip ATmega2560 is equipped with eleven, 8–bit general purpose, digital input/output (I/O) ports designated:

- PORTA (8 bits, PORTA[7:0])
- PORTB (8 bits, PORTB[7:0])
- PORTC (7 bits, PORTC[7:0])
- PORTD (8 bits, PORTD[7:0])
- PORTE (8 bits, PORTE[7:0])
- PORTF (8 bits, PORTF[7:0])
- PORTG (7 bits, PORTG[7:0])
- PORTH (8 bits, PORTH[7:0])
- PORTJ (8 bits, PORTJ[7:0])
- PORTK (8 bits, PORTK[7:0])
- PORTL (7 bits, PORTL[7:0])

All of the ports also have alternate functions which will be described later. In this section, we concentrate on the basic digital I/O port features.

As shown in Fig. 1.21, each port has three registers associated with it:

- Data Register PORTx—used to write output data to the port,
- Data Direction Register DDRx—used to set a specific port pin to either output (1) or input (0), and
- Input Pin Address PINx—used to read input data from the port.

534069_1_En_1_Chapter-print ☑ TYPESET ☐ DISK ☐ LE ☑ CP Disp.:2/9/2022 Pages: 230 Layout: German_T5

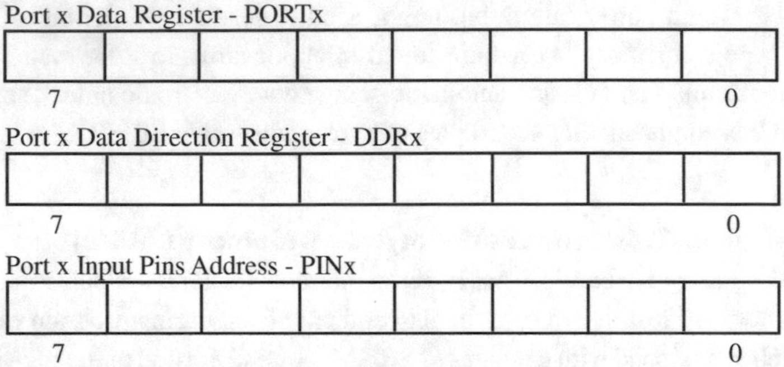

Port x Data Register - PORTx

7   0

Port x Data Direction Register - DDRx

7   0

Port x Input Pins Address - PINx

7   0

a) port associated registers

DDxn	PORTxn	I/O	Comment	Pullup
0	0	input	Tri-state (Hi-Z)	No
0	1	input	source current if externally pulled low	Yes
1	0	output	Output Low (Sink)	No
1	1	output	Output High (Source)	No

x: port designator (B, C, D)
n: pin designator (0 - 7)

b) port pin configuration

Fig. 1.21 ATmega2560 port configuration registers

Figure 1.21b describes the settings required to configure a specific port pin to either input or output. If selected for input, the pin may be selected for either an input pin or to operate in the high impedance (Hi–Z) mode. If selected for output, the pin may be further configured for either logic low or logic high.

Port pins are usually configured at the beginning of a program for either input or output and their initial values are then set. Usually all eight pins for a given port are configured simultaneously.

1.10.4 ATmega2560 Internal Systems

In this section, we provide a brief overview of the internal features of the ATmega2560. It should be emphasized that these features are the internal systems contained within the confines of the microcontroller chip. These built–in features allow complex and sophisticated tasks to be accomplished by the microcontroller.

534069_1_En_1_Chapter-print ☑ TYPESET ☐ DISK ☐ LE ☑ CP Disp.:2/9/2022 Pages: 230 Layout: German_T5

1.10.4.1 ATmega2560 Time Base

The microcontroller is a complex synchronous state machine. It responds to program steps in a sequential manner as dictated by a user–written program. The microcontroller sequences through a predictable fetch–decode–execute sequence. Each unique assembly language program instruction issues a series of signals to control the microcontroller hardware to accomplish instruction related operations.

The speed at which a microcontroller sequences through these actions is controlled by a precise time base called the clock. The clock source is routed throughout the microcontroller to provide a time base for all peripheral subsystems. The ATmega2560 may be clocked internally using a user–selectable resistor capacitor (RC) time base or it may be clocked externally. The RC internal time base is selected using programmable fuse bits. You may choose an internal fixed clock operating frequency of 128 kHz or 8 MHz. The clock frequency may be prescaled by a number of different clock division factors (1, 2, 4, etc.).

To provide for a wider range of frequency selections an external time source may be used. The external time sources, in order of increasing accuracy and stability, are an external RC network, a ceramic resonator, or a crystal oscillator. The system designer chooses the time base frequency and clock source device appropriate for the application at hand. Generally speaking, if the microcontroller will be interfaced to external peripheral devices either a ceramic resonator or a crystal oscillator should be used as a time base.

1.10.4.2 ATmega2560 Timing Subsystem

The ATmega2560 is equipped with a complement of timers which allows the user to generate a precision output signal, measure the characteristics (period, duty cycle, frequency) of an incoming digital signal, or count external events. Specifically, the ATmega2560 is equipped with two 8–bit timer/counters and four 16–bit timer/counters.

1.10.4.3 Pulse Width Modulation Channels

A pulse width modulated or PWM signal is characterized by a fixed frequency and a varying duty cycle. Duty cycle is the percentage of time a repetitive signal is logic high during the signal period. It may be formally expressed as:

$$duty\ cycle[\%] = (on\ time/period) \times (100\%)$$

The ATmega2560 is equipped with four 8–bit pulse width modulation (PWM) channels and 12 PWM channels with programmable resolution. The PWM channels coupled with the flexibility of dividing the time base down to different PWM subsystem clock source frequencies allow the user to generate a wide variety of PWM signals: from relatively high frequency low duty cycle signals to relatively low frequency high duty cycle signals.

PWM signals are used in a wide variety of applications including controlling the position of a servo motor and controlling the speed of a DC motor.

1.10.4.4 ATmega2560 Serial Communications

The ATmega2560 is equipped with a host of different serial communication subsystems including the Universal Synchronous and Asynchronous Serial Receiver and Transmitter (USART), the serial peripheral interface (SPI), and the Two–wire Serial Interface. What all of these systems have in common is the serial transmission of data. In a serial communications transmission, serial data is sent a single bit at a time from transmitter to receiver.

ATmega2560 Serial USART The serial USART may be used for full duplex (two way) communication between a receiver and transmitter. This is accomplished by equipping the ATmega2560 with independent hardware for the transmitter and receiver. The USART is typically used for asynchronous communication. That is, there is not a common clock between the transmitter and receiver to keep them synchronized with one another. To maintain synchronization between the transmitter and receiver, framing start and stop bits are used at the beginning and end of each data byte in a transmission sequence.

The ATmega2560 USART is quite flexible. It has the capability to be set to a variety of data transmission rates known as the Baud (bits per second) rate. The USART may also be set for data bit widths of 5–9 bits with one or two stop bits. Furthermore, the ATmega2560 is equipped with a hardware generated parity bit (even or odd) and parity check hardware at the receiver. A single parity bit allows for the detection of a single bit error within a byte of data. The USART may also be configured to operate in a synchronous mode.

ATmega2560 Serial Peripheral Interface–SPI The ATmega2560 Serial Peripheral Interface (SPI) can also be used for two–way serial communication between a transmitter and a receiver. In the SPI system, the transmitter and receiver share a common clock source. This requires an additional clock line between the transmitter and receiver but allows for higher data transmission rates as compared to the USART.

The SPI may be viewed as a synchronous 16–bit shift register with an 8–bit half residing in the transmitter and the other 8–bit half residing in the receiver. The transmitter is designated the master since it is providing the synchronizing clock source between the transmitter and the receiver. The receiver is designated as the slave.

ATmega2560 Two–wire Serial Interface–TWI The TWI subsystem allows the system designer to network related devices (microcontrollers, transducers, displays, memory storage, etc.) together into a system using a two–wire interconnecting scheme. The TWI allows a maximum of 128 devices to be interconnected together. Each device has its own unique address and may both transmit and receive over the two– wire bus at frequencies up to 400 kHz. This allows the device to freely exchange information with other devices in the network within a small area.

1.10.4.5 ATmega2560 Analog to Digital Converter–ADC

The ATmega2560 is equipped with a 16–channel analog to digital converter (ADC) subsystem. The ADC converts an analog signal from the outside world into a binary representation suitable for use by the microcontroller. The ATmega2560 ADC has 10–bit resolution. This means that an analog voltage between 0 and 5 V will be encoded into one of 1024 binary representations between $(000)_{16}$ and $(3FF)_{16}$. This provides the ATmega2560 with a voltage resolution of approximately 4.88 mV.

1.10.4.6 ATmega2560 Interrupts

The normal execution of a program follows a designated sequence of instructions. However, sometimes this normal sequence of events must be interrupted to respond to high priority faults and status both inside and outside the microcontroller. When these higher priority events occur, the microcontroller suspends normal operation and executes event specific actions contained within an interrupt service routine (ISR). Once the higher priority event has been serviced via the ISR, the microcontroller returns and continues processing the normal program.

The ATmega2560 is equipped with a complement of 57 interrupt sources. Eight interrupts are provided for external interrupt sources. Also, the ATmega2560 is equipped with three pin change interrupts. The remaining interrupts support the efficient operation of peripheral subsystems aboard the microcontroller.

1.11   Arduino Mega 2560 Open Source Schematic

The entire line of Arduino products is based on the visionary concept of open source hardware and software. That is, hardware and software developments are openly shared among users to stimulate new ideas and advance the Arduino concept. In keeping with the Arduino concept, the Arduino team openly shares the schematic of the Arduino Mega 2560 processing board. It is available for download at www.arduino.cc [1]. We employ the Mega 2560 R3 in several applications throughout the remainder of the book.

1.12   Extending the Hardware Features of the Arduino Platform

Additional features and external hardware may be added to selected Arduino platforms by using a daughter card concept. The daughter card is called an Arduino Shield as shown in Fig. 1.22. The shield mates with the header pins on the Arduino board. The shield provides a small fabrication area, a processor reset button, and a general use pushbutton and two light emitting diodes (LEDs). A large number of shields have been developed to provide extended specific features (e.g. motor control, communications, etc.) for the Arduino boards.

Fig. 1.22 Arduino shield.
Image used with permission
from SparkFun Electronics
(CC BY–NC–SA) (www.
sparkfun.com) [4]

1.13   Application: Dagu Rover 5 Robot

We equip the Dagu Rover 5 Robot for control by an Arduino UNO R3 as a maze navigating
robot. The Dagu Rover 5 robots is available in three variants:

- ROV5–1 is a tracked robot equipped with two motors and no motor encoders
- ROV5–2 is a tracked robot equipped with two motors and two motor encoders
- ROV5–3 is a tracked robot equipped with four motors and four motor encoders

In this example, we use the ROV5–1 platform. We replace the robot tracks with Dagu
28 mm wheels equipped with 4 mm hubs as shown in Fig. 1.23. It is controlled by two 7.2
VDC motors which independently drive a left and right wheel.

We equip the Rover 5 platform with three Sharp GP2Y0A21YKOF infrared (IR) sensors
as shown in Fig. 1.25. The sensors are available from SparkFun Electronics (www.sparkfun.
com) [4]. We mount the sensors on a bracket constructed from thin aluminum. The bracket is
attached to the Pololu RP5/Rover 5 extension plate (#1530) [www.pololu.com] (Fig. 1.24).

Dimensions for the bracket are provided in the figure. Alternatively, the IR sensors may
be mounted to the robot platform using "L" brackets available from a local hardware store.
The characteristics of the sensor are provided in Fig. 1.24. The robot is placed in a maze
with reflective walls. The project goal is for the robot to detect wall placement and navigate
through the maze. It is important to note the robot is not provided any information about the
maze. The control algorithm for the robot is hosted on the Arduino UNO R3.

Dagu 28 mm wheels
with 4 mm hub

Fig. 1.23 Dagu Rover 5 Robot with wheels (www.dagurobot.com)

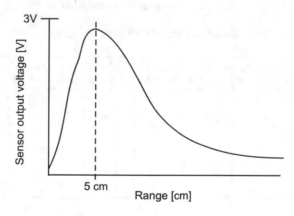

Fig. 1.24 Sharp GP2Y0A21YKOF IR sensor profile [3]

1.13.1 Requirements

The requirements for this project are simple: the robot must autonomously navigate through the maze without touching maze walls. With the interface circuit employed, the robot motors may only be moved in the forward direction. To render a left turn, the left motor is stopped and the right motor is asserted until the robot completes the turn. To render a right turn, the

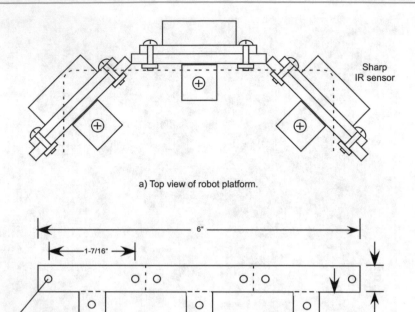

a) Top view of robot platform.

Sharp
IR sensor

b) Construction details for sensor bracket.

Fig. 1.25 Dagu Rover 5 robot platform modified with three IR sensors

	Left Sensor	Middle Sensor	Right Sensor	Wall Left	Wall Middle	Wall Right	Left Motor	Right Motor	Left Signal	Right Signal	Comments
0	0	0	0	0	0	0	1	1	0	0	Forward
1	0	0	1	0	0	1	1	1	0	0	Forward
2	0	1	0	0	1	0	1	0	0	1	Right
3	0	1	1	0	1	1	0	1	1	0	Left
4	1	0	0	1	0	0	1	1	0	0	Forward
5	1	0	1	1	0	1	1	1	0	0	Forward
6	1	1	0	1	1	0	1	0	0	1	Right
7	1	1	1	1	1	1	1	0	0	1	Right

Fig. 1.26 Truth table for robot action

opposite action is required. To complete the robot truth table provided in Fig. 1.26, imagine you are the robot. What would you do to avoid a maze wall when walls are sensed as described in each row of the table?

The task in writing the control algorithm is to take the UML activity diagram provided in Fig. 1.29 and the actions specified in the robot action truth table and transform both into a coded algorithm. This may seem formidable but we take it a step at a time.

534069_1_En_1_Chapter-print ☑ TYPESET ☐ DISK ☐ LE ☑ CP Disp.:2/9/2022 Pages: 230 Layout: German_T5

1.13.2 Circuit Diagram–Arduino UNO

The circuit diagram for the robot is provided in Fig. 1.27. The three IR sensors (left, middle, and right) are mounted on the leading edge of the robot to detect maze walls. The output from the sensors is fed to three Arduino UNO R3 ADC channels (ANALOG IN 0–2). The robot motors are driven by PWM channels (PWM: DIGITAL 11 and PWM: DIGITAL 10). The Arduino UNO R3 is interfaced to the motors via a Darlington NPN transistor (TIP120). The transistor has enough drive capability to handle the maximum current requirements of the motor. Since the motor power supply is at 9 VDC and the motors are rated at 7.2 VDC, three 1N4001 diodes are placed in series with the motor. This reduces the supply voltage to the motor to be approximately 6.9 VDC. The robot is powered by a 9 VDC power supply which is fed to a 5 VDC voltage regulator. To save on battery expense, it is recommended to use a 9 VDC, 2A rated inexpensive, wall–mount power supply. A power umbilical of braided wire may be used to provide power to the robot while navigating about the maze. We explore portable battery supplies in a upcoming chapter.

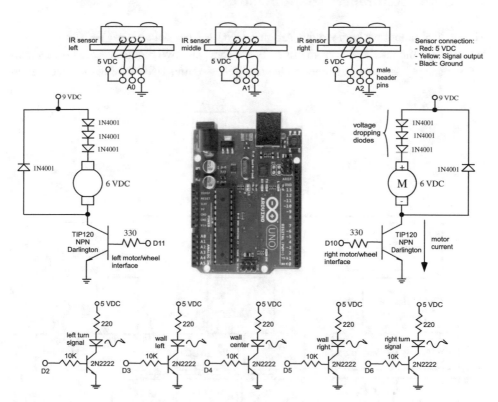

Fig. 1.27 Robot circuit diagram. UNO R3 illustration used with permission of the Arduino Team (CC BY–NC–SA) (www.arduino.cc)

534069_1_En_1_Chapter-print ☑ TYPESET ☐ DISK ☐ LE ☑ CP Disp.:2/9/2022 Pages: 230 Layout: German_T5

1.13.3 Structure Chart

A structure chart diagrams how required actions will be partitioned among different functions. Directional arrows are used to indicate the flow of data variables in and out of the function. It is a powerful tool in getting the big picture of a program. The structure chart for the robot project is provided in Fig. 1.28.

1.13.4 UML Activity Diagrams

A UML activity diagram is a Unified Modeling Language compliant flow chart. It diagrams program flow. The UML activity diagram for the robot is provided in Fig. 1.29. The structure chart and UML activity diagram provides a map for constructing a complex program.

1.13.5 Microcontroller Code–Arduino UNO

The control algorithm begins with Arduino UNO R3 pin definitions. Variables are then declared for the readings from the three IR sensors. The two required Arduino functions follow: setup() and loop(). In the setup() function, Arduino UNO R3 pins are declared as

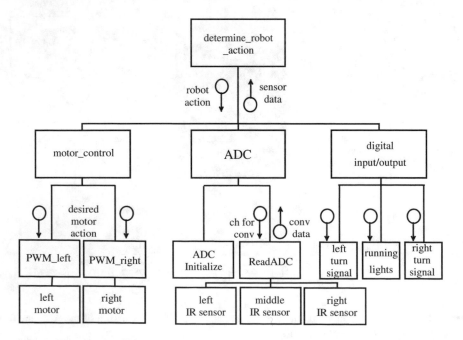

Fig. 1.28 Robot structure diagram

534069_1_En_1_Chapter-print ☑TYPESET ☐DISK ☐LE ☑CP Disp.:2/9/2022 Pages: 230 Layout: German_T5

Fig. 1.29 Robot UML activity
diagram

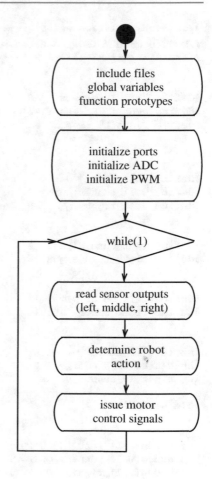

output. The loop() begins by reading the current value of the three IR sensors. The 512 threshold value corresponds to a desired IR sensor range. This value may be adjusted to change the range at which the maze wall is detected. The IR sensor readings are followed by an eight part if–else if statement. The statement contains a part for each row of the truth table provided in Fig. 1.26. For a given configuration of sensed walls, the appropriate wall detection LEDs are illuminated followed by commands to activate the motors (analogWrite) and illuminate the appropriate turn signals. The analogWrite command issues a signal from 0 to 5 VDC by sending a constant from 0 to 255 using pulse width modulation (PWM) techniques. The turn signal commands provide two actions: the appropriate turns signals are flashed and a 1.5 s total delay is provided. This provides the robot 1.5 s to render a turn. This delay may need to be adjusted during the testing phase.

```
//***********************************************************************
//Dagu 5 Maze Following Robot
//***********************************************************************

                                    //analog input pins
#define left_IR_sensor    A0        //analog pin - left IR sensor
#define center_IR_sensor A1         //analog pin - center IR sensor
#define right_IR_sensor  A2         //analog pin - right IR sensor

                                    //digital output pins
                                    //LED indicators - wall detectors
#define wall_left         3         //digital pin - wall_left
#define wall_center       4         //digital pin - wall_center
#define wall_right        5         //digital pin - wall_right

                                    //LED indicators - turn signals
#define left_turn_signal  2         //digital pin - left_turn_signal
#define right_turn_signal 6         //digital pin - right_turn_signal

                                    //motor outputs
#define left_motor        11        //digital pin - left_motor
#define right_motor       10        //digital pin - right_motor


int left_IR_sensor_value;           //variable for left IR sensor
int center_IR_sensor_value;         //variable for center IR sensor
int right_IR_sensor_value;          //variable for right IR sensor

void setup()
  {
                                    //LED indicators - wall detectors
  pinMode(wall_left,    OUTPUT);    //configure pin 1 for digital output
  pinMode(wall_center,  OUTPUT);    //configure pin 2 for digital output
  pinMode(wall_right,   OUTPUT);    //configure pin 3 for digital output

                                    //LED indicators - turn signals
  pinMode(left_turn_signal,OUTPUT); //configure pin 0 for digital output
  pinMode(right_turn_signal,OUTPUT);//configure pin 4 for digital output

                                    //motor outputs - PWM
  pinMode(left_motor,   OUTPUT);    //config pin 11 for digital output
  pinMode(right_motor,  OUTPUT);    //config pin 10 for digital output
  }

void loop()
  {
                                    //read analog output from IR sensors
  left_IR_sensor_value   = analogRead(left_IR_sensor);
  center_IR_sensor_value = analogRead(center_IR_sensor);
  right_IR_sensor_value  = analogRead(right_IR_sensor);

  //robot action table row 0
  if((left_IR_sensor_value < 512)&&(center_IR_sensor_value < 512)&&
```

```
    (right_IR_sensor_value < 512))
    {
                                            //wall detection LEDs
    digitalWrite(wall_left,    LOW);        //turn LED off
    digitalWrite(wall_center, LOW);         //turn LED off
    digitalWrite(wall_right,   LOW);        //turn LED off
                                            //motor control
    analogWrite(left_motor,   128);         //0 (off) to
                                            //255 (full speed)
    analogWrite(right_motor, 128);          //0 (off) to
                                            //255 (full speed)
                                            //turn signals
    digitalWrite(left_turn_signal,  LOW);   //turn LED off
    digitalWrite(right_turn_signal, LOW);   //turn LED off
    delay(500);                             //delay 500 ms
    digitalWrite(left_turn_signal,  LOW);   //turn LED off
    digitalWrite(right_turn_signal, LOW);   //turn LED off
    delay(500);                             //delay 500 ms
    digitalWrite(left_turn_signal,  LOW);   //turn LED off
    digitalWrite(right_turn_signal, LOW);   //turn LED off
    delay(500);                             //delay 500 ms
    digitalWrite(left_turn_signal,  LOW);   //turn LED off
    digitalWrite(right_turn_signal, LOW);   //turn LED off
    analogWrite(left_motor, 0);             //turn motor off
    analogWrite(right_motor,0);             //turn motor off
    }

//robot action table row 1
else if((left_IR_sensor_value < 512)&&(center_IR_sensor_value < 512)&&
        (right_IR_sensor_value > 512))
    {
                                            //wall detection LEDs
    digitalWrite(wall_left,    LOW);        //turn LED off
    digitalWrite(wall_center, LOW);         //turn LED off
    digitalWrite(wall_right,   HIGH);       //turn LED on
                                            //motor control
    analogWrite(left_motor,   128);         //0 (off) to
                                            //255 (full speed)
    analogWrite(right_motor, 128);          //0 (off) to
                                            //255 (full speed)
                                            //turn signals
    digitalWrite(left_turn_signal,  LOW);   //turn LED off
    digitalWrite(right_turn_signal, LOW);   //turn LED off
    delay(500);                             //delay 500 ms
    digitalWrite(left_turn_signal,  LOW);   //turn LED off
    digitalWrite(right_turn_signal, LOW);   //turn LED off
    delay(500);                             //delay 500 ms
    digitalWrite(left_turn_signal,  LOW);   //turn LED off
    digitalWrite(right_turn_signal, LOW);   //turn LED off
    delay(500);                             //delay 500 ms
    digitalWrite(left_turn_signal,  LOW);   //turn LED off
    digitalWrite(right_turn_signal, LOW);   //turn LED off
    analogWrite(left_motor, 0);             //turn motor off
```

534069_1_En_1_Chapter-print ☑ TYPESET ☐ DISK ☐ LE ☑ CP Disp.:2/9/2022 Pages: 230 Layout: German_T5

```
        analogWrite(right_motor,0);              //turn motor off
        }


        :
        :  //Insert code for rows 2 - 7
        :

    //robot action table row 7
    else if((left_IR_sensor_value > 512)&&(center_IR_sensor_value > 512)&&
            (right_IR_sensor_value > 512))
        {
                                                 //wall detection LEDs
        digitalWrite(wall_left,    HIGH);        //turn LED on
        digitalWrite(wall_center, HIGH);         //turn LED on
        digitalWrite(wall_right,   HIGH);        //turn LED on
                                                 //motor control
        analogWrite(left_motor,   128);          //0 (off) to
                                                 //255 (full speed)
        analogWrite(right_motor, 0);             //0 (off) to
                                                 //255 (full speed)
                                                 //turn signals
        digitalWrite(left_turn_signal,  LOW);    //turn LED off
        digitalWrite(right_turn_signal, HIGH);   //turn LED on
        delay(500);                              //delay 500 ms
        digitalWrite(left_turn_signal,  LOW);    //turn LED off
        digitalWrite(right_turn_signal, LOW);    //turn LED off
        delay(500);                              //delay 500 ms
        digitalWrite(left_turn_signal,  LOW);    //turn LED off
        digitalWrite(right_turn_signal, HIGH);   //turn LED on
        delay(500);                              //delay 500 ms
        digitalWrite(left_turn_signal,  LOW);    //turn LED off
        digitalWrite(right_turn_signal, LOW);    //turn LED off
        analogWrite(left_motor, 0);              //turn motor off
        analogWrite(right_motor,0);              //turn motor off
        }
}


//************************************************************************
```

1.14   Application: Tinkerkit Braccio

The Arduino Tinkerkit Braccio is a six joint robotic arm. Each joint is controlled by a servo motor. The kit comes equipped with Braccio shield to provide for interface between the Arduino UNO R3 and the robotic arm as shown in Fig. 1.30. The angular extent of each joint is provided in the figure.

There are several Braccio support libraries available for download via the Arduino Development Environment. The sketch "simpleMovements," included with the library demonstrates the extension of each robot arm joint.

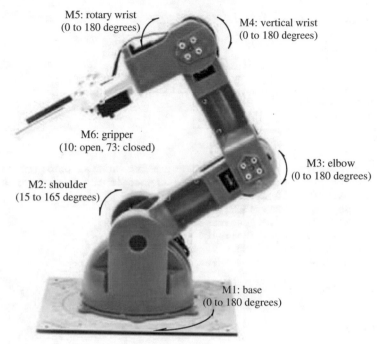

M5: rotary wrist
(0 to 180 degrees)

M4: vertical wrist
(0 to 180 degrees)

M6: gripper
(10: open, 73: closed)

M3: elbow
(0 to 180 degrees)

M2: shoulder
(15 to 165 degrees)

M1: base
(0 to 180 degrees)

TinkerKit Braccio Robotic Arm

Tinkerkit Braccio Shield

Joint Summary	
M1: base	0 to 180 degrees
M2: shoulder	15 to 165 degrees
M3: elbow	0 to 180 degrees
M4: vertical wrist	0 to 180 degrees
M5: rotary wrist	0 to 180 degrees
M6: gripper	10: open
	73: closed

Fig. 1.30 Arduino Tinkerkit Braccio six joint robotic arm. Arduino Braccio illustrations used with permission of the Arduino Team (CC BY–NC–SA) (www.arduino.cc) [5]

```
//****************************************************************
//simpleMovements.ino
//
//This  sketch simpleMovements shows how they move each servo motor
//of Braccio.
//
//Created on 18 Nov 2015 by Andrea Martino.
//This example is in the public domain.
//****************************************************************

#include <Braccio.h>
```

```
#include <Servo.h>

Servo base;
Servo shoulder;
Servo elbow;
Servo wrist_rot;
Servo wrist_ver;
Servo gripper;

void setup()
{
//Initialization functions and set up the initial position for Braccio
//All the servo motors will be positioned in the "safety" position:
//   - Base (M1):90 degrees
//   - Shoulder (M2): 45 degrees
//   - Elbow (M3): 180 degrees
//   - Wrist vertical (M4): 180 degrees
//   - Wrist rotation (M5): 90 degrees
//   - gripper (M6): 10 degrees

Braccio.begin();
}

void loop()
{
//Step Delay: a milliseconds delay between the movement of each servo.
//Allowed values from 10 to 30\,ms.
//   M1 = base degrees. Allowed values from 0 to 180 degrees
//   M2 = shoulder degrees. Allowed values from 15 to 165 degrees
//   M3 = elbow degrees. Allowed values from 0 to 180 degrees
//   M4 = wrist vertical degrees. Allowed values from 0 to 180 degrees
//   M5 = wrist rotation degrees. Allowed values from 0 to 180 degrees
//   M6 = gripper degrees. Allowed values from 10 to 73 degrees.
//        10: the gripper is open, 73: the gripper is closed.


                    //(step delay,  M1,  M2,  M3,  M4, M5,  M6);
Braccio.ServoMovement(20,            0,  15, 180, 170,  0,  73);
delay(1000);        //wait 1 second
Braccio.ServoMovement(20,          180, 165,   0,  0,180,  10);
delay(1000);        //wait 1 second
}

//********************************************************************
```

Exercise: Modify the "simpleMovements" sketch to demonstrate the maximum range of each joint sequentially.

534069_1_En_1_Chapter-print ☑ TYPESET ☐ DISK ☐ LE ☑ CP Disp.:2/9/2022 Pages: 230 Layout: German_T5

1.15 Summary

The goal of this chapter was to provide a tutorial on how to begin programming a sketch with the Arduino Development Environment. We used a top–down design approach. We began with the "big picture" of the chapter followed by an overview of the Arduino Development Environment. We then provided an overview of the Arduino concept of open source hardware. This was followed by a description of the Arduino UNO R3 processor board powered by the ATmega328. An overview of ATmega2560 systems followed. Throughout the chapter, we provided examples and also provided references to a number of excellent references.

1.16 Problems

1. Describe the steps in writing a sketch and executing it on an Arduino UNO R3 processing board.
2. What is the serial monitor feature used for in the Arduino Development Environment?
3. Describe what variables are required and returned and the basic function of the following built–in Arduino functions: Blink, Analog Input.
4. Describe in your own words the Arduino open source concept.
5. Sketch a block diagram of the ATmega328 and its associated systems. Describe the function of each system.
6. Describe the different types of memory components within the ATmega328. Describe applications for each memory type.
7. Describe the three different register types associated with each port.
8. How may the features of the Arduino UNO R3 be extended?
9. Describe the required features of an external power supply used with an Arduino UNO R3.
10. What is the primary purpose of an interrupt? Why is this important?
11. We equipped the Dagu Rover R5 robot with three Sharp GP2Y0A21YK0F Distance Measuring Sensor Units. Carefully profile the response of the IR sensor. Provide a voltage output versus range plot for the sensor. What is the angular response profile of the sensor? With the data collected, develop a sketch of the sensor detection lobe.
12. Using the data collected in the question above, design a bracket to hold three of the Sharp GP2Y0A21YK0F Distance Measuring Sensor Units to minimize blinds spots.
13. The Sharp GP2Y0A21YK0F Distance Measuring Sensor Unit is an IR sensor. How may the sensor be shielded to prevent interference from other IR light sources not emanating from the Sharp sensor? **Hint:** Why do we wear ball caps outside? Why do we equip horses with blinders?
14. Develop a table of available Arduino UNO R3 compatible shields that may be useful in a DIY robot project.

534069_1_En_1_Chapter-print ☑ TYPESET ☐ DISK ☐ LE ☑ CP Disp.:2/9/2022 Pages: 230 Layout: German_T5

15. In the robot control algorithm provided in the chapter, we left the motors run for 1500 ms (1.5 s) each time a new action was initiated. How far did the robot move for each action? How can we be sure?

16. Summarize the differences between the Arduino UNO R3 and Mega 2560. How would you choose between the two in a given application?

References

1. Arduino homepage, www.arduino.cc.
2. *Microchip 8–bit AVR Microcontroller with 4/8/16/32 K Bytes In–System Programmable Flash, ATmega48PA, 88PA, 168PA, 328P* data sheet: 8171D–AVR–05/11, Microchip Corporation, 2325 Orchard Parkway, San Jose, CA 95131.
3. *Sharp GP2Y0A21YK0F Distance Measuring Sensor Unit*, Sheet No. E4–A00201EN, Sharp Corporation, 2006.
4. SparkFun Electronics, 6175 Longbow Drive, Suite 200, Boulder, CO 80301 (www.sparkfun.com).
5. *Tinkerkit Braccio Robot*, https://store.arduino.cc/usa/tinkerkit-braccio.

Introduction to Low-Cost 3D Printing

2

Objectives: After reading this chapter, the reader should be able to do the following:

- Describe how 3D printing works, what the most common categories of 3D printing are, the pros and cons of some of the more popular types of 3D printing, as well as some exciting current use–cases for 3D printing across a wide range of industries;
- Describe common 3D printing anatomy, specifically how fused deposition modeling (FDM) 3D printing works;
- List affordable and popular brands of desktop FDM 3D printers;
- List and describe popular materials used in FDM 3D printing;
- Describe how to use the common slicer software to prepare a model for printing;
- Summarize tips and tricks for operating FDM printers, including what to do before, during, and after your project;
- Describe common parameters and settings to consider, including support, infill, and build plate adhesion; and
- Describe how to start 3D printing parts for any type of Arduino project.

© The Author(s), under exclusive license to Springer Nature Switzerland AG 2022
T. Kerr and S. Barrett, *Arduino IV: DIY Robots*, Synthesis Lectures on Digital Circuits
& Systems, https://doi.org/10.1007/978-3-031-11209-6_2

2.1   3D Printing 101

2.1.1   Overview

You may have seen 3d printing at school, in your local library or community center, or even on tv. It's an extraordinary technology used across an ever–growing list of industries, from movies and museums to mechanical engineering and the Met Gala. Today, you are very likely to see 3D printers in makerspaces, libraries, schools, workshops, dentist's offices, Michelin–starred restaurants, automotive assembly lines, hospitals, operating rooms, manufacturing facilities, construction sites, farms, jewelry shops, and even the space station! This chapter presents a general overview of how 3D printing works and what types of projects you might be able to make with the more popular desktop 3D printers on the market. You'll get a comprehensive guide on how to start 3D printing, as well as the tips and tricks to become a 3D printing pro. In this chapter, we're going to review a wide variety of different types of 3D printing technologies, but will focus primarily on the most popular and widely accessible 3D print technology on the market: fused deposition modeling (FDM) 3D printing. When we discuss 3D printing in broad strokes, know that we're talking about FDM machines. If you're interested in integrating 3D printing into your robot projects, this is the place to start.

2.1.2   What Is 3D Printing?

In a very basic sense, you might think of FDM 3D printing like using a computer–controlled hot glue gun. Just like a hot glue gun, 3D printers heat a cylinder of material and eject it out a hot nozzle. However, there are a few key distinctions that set the two technologies apart. Unlike actual hot glue guns, our "super–smart hot glue gun" 3D printer has a very tiny nozzle averaging only 400 μm (0.4 mm) in diameter. And instead of squeezing out a cylinder of hot glue, FDM 3D printers eject a thin string of molten "thermoplastic," which is simply a plastic designed to melt at a specific temperature (usually somewhere around 200–250 °C). There are typically two components of a 3D printer working together to build 3D shapes on the build plate. One component uses motors to grip and drive the thermoplastic "filament" forward through the nozzle at a regular rate. At the same time, another part of the 3D printer heats up the filament to a set temperature to allow it to ooze out of the nozzle at a set width and height. Working together, these components allow molten plastic to flow or "extrude" out of a tiny nozzle at a regular and predictable rate. Once extruded, fans quickly cool the molten plastic in place as the computer pilots the nozzle to different print areas on the build plate. In such a way, a 3D printer can ultimately build a 3D shape up out of nothing by printing thin lines of thermoplastic layer–by–layer. Every 3D printer builds parts based on the same basic idea: a digital 3D model is turned into a physical 3D object by adding or fusing material a layer at a time. This is where the term "additive manufacturing" comes from. At its core, 3D printing is a very different method of producing parts compared to traditional subtractive manufacturing like CNC machining, laser cutting, or formative manufacturing like injection molding. It's not too big a leap to suggest that a fourth industrial revolution driven by additive manufacturing is on the horizon. As traditional production–manufacturing shifts, we're on the cusp of a "desktop manufacturing revolution" because of technologies like 3D printing and rapid prototyping. Some 20 or 30 years ago, 3D printers may have only been accessible to advanced research labs. Now, those 3D printers are faster, easier to use, more affordable, and can fit in makerspaces, garages, libraries, workshops, homes, retail stores, and space stations. Because this type of technology is so much more accessible, 3D printing can lower costs, save time, and go far beyond the limits of traditional fabrication processes for product development. From concept models and functional prototypes to jigs, fixtures, specialized robot parts, or even end–use parts in manufacturing, 3D printing technologies offer versatile solutions in a wide variety of applications and in all sorts of shapes and sizes. And that means that 3D printers are revolutionizing how we can make almost anything.

2.1.3   Common Categories of 3D Printing

3D printers might seem daunting at first, but they can be surprisingly easy to use, even if you don't consider yourself a particularly tech–savvy person. Gone are the days when you might have to write thousands of lines of code or program the machine's every X, Y, and Z movement. Today you can drag–and–drop 3D models into 3D printing software, choose a handful of settings, load your material, and hit "Print." To start, you'll need to have a handle on what your project's end goals are. Are you creating a hobby project that will sit on your desk? Will your project bear weight? Should it be waterproof or airtight? Does it need to look photorealistic? Will it need to have a high dimensional accuracy? These choices will influence what type of 3D printer you want to use.

With this in mind, the first speedbump newcomers run into with 3D printing technology is the types of 3D printers and materials that they should use. What does FDM and SLA mean? What about SLS, EBM, and DMLS? Or MJ and DOD? The acronyms can get confusing fairly quickly, but stick with us—that's what this chapter is about.

The first thing to understand is that 3D printing is a relatively general term covering seven major additive manufacturing categories. Within those seven categories, there are 18 different types of 3D printing technology. Every day, clever folks from all around the world develop new techniques and technologies, so the list will surely grow. At the time of this book's publication, the seven categories of 3D printing and 18 types of 3D printing technologies include:

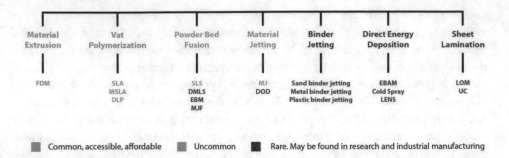

- Material Extrusion
  - Fused Deposition Modeling (FDM)

- Vat Polymerization
  - Stereolithography (SLA)
  - Masked Stereolithography (MSLA)
  - Digital Light Processing (DLP)

- Powder Bed Fusion
  - Selective Laser Sintering (SLS)
  - Direct Metal Laser Sintering (DMLS)
  - Electron Beam Melting (EBM)
  - Multi Jet Fusion (MJF)

- Material Jetting
  - Material Jetting (MJ)
  - Drop on Demand (DOD)

- Binder Jetting
  - Sand Binder Jetting
  - Metal Binder Jetting
  - Plastic Binder Jetting

- Direct Energy Deposition
  - Electron Beam Additive Manufacturing (EBAM)
  - Cold Spray
  - Laser Engraved Net Shaping (LENS)

- Sheet Lamination
  - Laminated Object Manufacturing (LOM)
  - Ultrasonic Consolidation (UC)

Rest assured, these won't be on the final test! Instead, we'll take a very brief look at the four most common types of 3D printing that you're most likely to bump into in places like makerspaces, workshops, libraries, manufacturing facilities, or schools. These include fused deposition modeling (FDM), stereolithography (SLA), selective laser sintering (SLS), and material jetting (MJ). Following this, we'll focus on FDM 3D printing for the remainder of the chapter, as it is undoubtedly the most common 3D printing method available.

2.1.3.1 Fused Deposition Modeling (FDM/FFF)

Fused deposition modeling, or "FDM" 3D printing (synonymous with another term: fused filament fabrication: "FFF"), is a popular method of additive manufacturing where layers of melted thermoplastic are fed through and extruded out a small hot nozzle and fused together in a pattern to create a 3D object. The material is usually heated up just past the temperature that it begins to become viscous and melt, and then extruded in a pattern next to or on top of previous extruded layers, creating an object layer by layer as shown in Fig. 2.1.

FDM is very accessible, user–friendly, and exceptionally well–suited for proof–of–concept prototypes, educational models, as well as quick and low–cost parts that might otherwise be machined more slowly. FDM can also print in a wide variety of strong, flexible, weather–resistant, hardy, or otherwise exotic materials. This includes materials like wood pulp to be sanded and stained, metal particulate to be polished and oxidized, chocolate to be molded into delicious geometric shapes, and even cement to build houses!

2.1.3.2 Stereolithography (SLA)

Stereolithography, or "SLA" 3D printing, holds the interesting distinction of being the world's first 3D printing technology, despite looking like something from a recent sci–fi film. Surprisingly, SLA was invented in 1986 by Chuck Hull, while in contrast, FDM was patented in 1992.

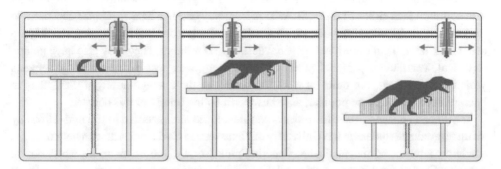

Fig. 2.1 The fused deposition modeling (FDM) process. A thermoplastic is ejected out a hot nozzle and cooled in place by fans

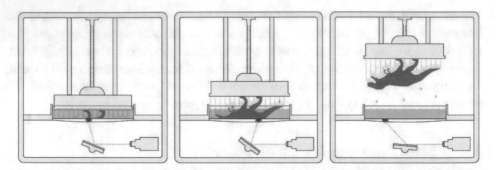

Fig. 2.2 The stereolithography (SLA) process. A liquid photopolymer is cured onto a build plate by a precise laser

An SLA printer uses special mirrors known as galvanometers that precisely angle and steer a laser beam across a vat of liquid photopolymer resin, which selectively cures and hardens an object layer by layer as the build plate and cured object are pulled out of the uncured resin goo slowly, as shown in Fig. 2.2. Think of this sort of like Han Solo coming out of carbonite.

SLA parts are known to have some of the highest resolution and accuracy, clearest details, and best surface finish of all desktop 3D printing technologies. Material manufacturers have created innovative SLA photopolymer resins with a huge range of different material properties tailor–made for engineering, industrial, medical, and research purposes. SLA is excellent for highly detailed prototypes requiring tight tolerances and smooth surfaces, such as molds, patterns, and functional parts. SLA is used in a variety of fields, including engineering, rapid prototyping, manufacturing, dentistry, jewelry, and education.

2.1.3.3 Selective Laser Sintering (SLS)

Selective laser sintering, or "SLS," is another popular type of 3D printing, though it's a bit less common than either FDM or SLA. This is largely because SLS machines use lasers to fuse a bed of fine–grained powder, which usually means they can be a bit messier, and they are often a bit larger than traditional personal desktop 3D printers. They may also have unique electrical, ventilation, personal protective equipment, and post–processing requirements which means they're not quite as easily accessible to the average hobbyist. Still, they're becoming more and more popular, and thus worth noting briefly in this chapter.

If you can believe it, the SLS process was developed and patented in the mid–1980s by an undergraduate student at the University of Texas named Carl Deckard, with help from his academic adviser and mechanical engineering professor, Dr. Joe Beaman. It's worth noting here that Deckard and Beaman patented the SLS 3D printing process but certainly were not the first to sinter objects. Sintering, the process of compacting and forming an object using heat or pressure without melting the object, has been used for thousands of years. Bricks,

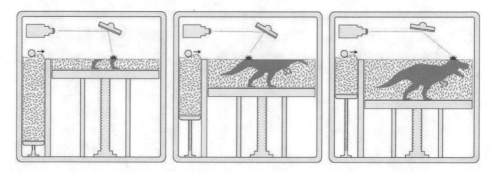

Fig. 2.3 The selective laser sintering (SLS) process. Thin layers of fine powdered material are selectively sintered, or fused, by a laser

metals, ceramics, and jewelry all make use of the sintering process. If you've ever packed a hard snowball by pressing loose snow together, you've practiced sintering!

Typically during the SLS process, tiny particles of powdered material like nylon plastic, ceramic, or glass are fused together by heat from a high–power laser that traces the cross–section of the design, as shown in Fig. 2.3. As each micron–thin layer of material is fused, the machine lowers the print bed down a tiny amount and gently pushes another micron–thin layer of material across the top of the previous layer, a bit like a miniature bulldozer. This new layer is subsequently sintered to the previous layer, and then the whole process repeats. Gradually, the 3D printer forms a sintered 3D object supported by the surrounding unsintered powder.

One of the most notable benefits of SLS over other types of 3D printing technologies is that the powder bed filled with unsintered powder surrounding the 3D object can support the object as it prints, which helps hold the entire thing together. As we'll learn later, FDM and SLA 3D printing both typically require any type of overhanging parts (like arches, bridges, or bits that stick out far from the model) to be supported by a latticework of material that is later thrown away or recycled. It's pretty common for FDM and SLA models to require supports, but SLS parts use the surrounding material to do that job instead. This means that despite being a bit messier, SLS parts require less post–processing or additional sanding to tidy up an object once that object is 3D printed.

2.1.3.4 Material Jetting (MJ)

Did you know you can 3D print photorealistic, full–color, fully articulated objects? Material Jetting ("MJ"), otherwise known as polyjet 3D printing, is arguably one of the most fascinating 3D printing processes available today, as shown in Fig. 2.4. MJ works by selectively depositing a rainbow–colored mist of incredibly tiny droplets of a liquid resin material onto a build plate, which is then cured in place by powerful ultraviolet (UV) lights. In a way, these 3D printers work a bit like a paper printer, though instead of mixing a single layer of

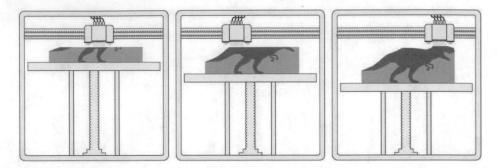

Fig. 2.4 The material jetting (MJ) process. Tiny droplets of liquid resin are jetted out of a print head and cured by UV light

cyan, magenta, yellow, and black (CYMK) ink, MJ printers spray incredibly small droplets of CMYK photopolymer resin. After one layer has been laid down and cured by UV light, the build platform is lowered down, and the process is repeated until a 3D object is formed.

Unlike other types of 3D printing that focus on a single point or area at a time and use a laser or a nozzle that follows a predefined cross–sectional path, MJ machines rapidly deposit larger amounts of material in a line. MJ printers typically zip back and forth, left and right across the print surface like the print head on a paper printer. As a result, MJ printers can be a bit speedier than traditional FDM or SLA printers, although the cost of materials is typically a bit higher. Even though these printers are much more complex than conventional FDM or SLA printers, they're becoming more affordable and increasingly common in colleges, universities, and libraries. MJ is best known as one of the only types of 3D printing technology capable of printing full–color objects made of multiple materials simultaneously. Meaning that, yes, you can 3D print an entire shoe.

2.1.4   Best Uses of 3D Printing

3D printing is used in a huge range of industries. To highlight the sheer versatility of 3D printing, it might be helpful to briefly explore some of the vast and varied 3D printing applications. For example, suppose you're looking to build the chassis for an Arduino robot, design a beautiful geodesic dome out of chocolate, or construct a Martian habitat to keep space explorers safe. All of these examples are possible with 3D printing.

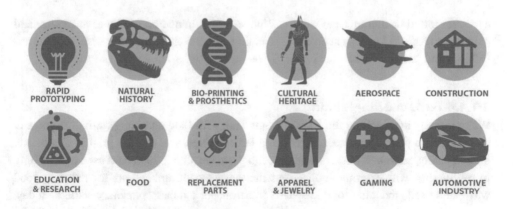

RAPID PROTOTYPING	NATURAL HISTORY	BIO-PRINTING & PROSTHETICS	CULTURAL HERITAGE	AEROSPACE	CONSTRUCTION
EDUCATION & RESEARCH	FOOD	REPLACEMENT PARTS	APPAREL & JEWELRY	GAMING	AUTOMOTIVE INDUSTRY

2.1.4.1 Rapid Prototyping, Manufacturing

With lower–cost desktop 3D printing more available to consumers, now anyone can make prototypes in an afternoon. For those with a ground–breaking idea, rapid prototyping and rapid manufacturing using 3D printers can allow someone to test out many different designs and alter or test the size, shape, tolerances, and iterative variations of a design at a very fast rate. Ultimately, rapid prototyping with 3D printers enables innovators to bring better products to market more quickly than ever before.

2.1.4.2 Education and Research

Researchers and educators can use 3D printing to complement existing curriculum or create custom parts, components, or tools to conduct new or niche research. A 3D printer might serve as Ms. Frizzle's Magic School Bus machine for educators. They can scale microscopic life such as tardigrades up and place the 3D printed models in peoples' hands, or scale things down from the size of galaxies and do the same. They enable educators in rural and remote communities to download and produce exciting hands–on lesson plans for their students. Three–dimensional printers can also allow students of all ages and abilities to get direct and interactive engagement with objects they might otherwise never be able to see, such as rare museum collections or comparative studies.

2.1.4.3 Art, Fashion, and Jewelry

With the ability to 3D print sneakers, dresses, beachwear, and even jewelry, 3D printing is increasingly popular with artists and designers around the world. Today, there are 3D printers that can print on cloth, 3D printers that can print in metal or castable metal mold materials such as wax or plastic, and 3D printers that can print rubber soles or custom orthotics. It's no wonder that today Nike and New Balance are both using 3D printing to produce custom–fitted shoes for athletes, or that designers are even able to make custom eye wear based on unique head shapes. With so many unique body shapes and sizes, and paired alongside technology such as 3D scanning, 3D printers seem tailor–made for the art, fashion, and jewelry world.

2.1.4.4 Food and Bio–Printing

We're quickly approaching the world of Star Trek with food and bio–printing. If you can make it into a paste and squeeze it out of a large syringe, you can 3D print it. Chocolate, dough, candy, pancakes, burgers, and broccoli paste are all fair game and used increasingly in restaurants, with exciting applications that have broad implications for reducing food waste and even producing food in space for astronauts and early planetary settlers. Today, companies are even exploring how to 3D print meat by cultivating stem cells to build fat and muscle!

Within the world of bio–printing, 3D printing functional human organs are also only a heartbeat away. Scientists and engineers are working on methods to 3D print a carefully placed latticework of stem cells to produce new tissues and organs for humans. In fact, scientists and engineers were recently able to 3D print the world's first beating human heart, complete with cells, blood vessels, ventricles, and chambers! While the prototype heart is small—only the size of a rabbit's heart—it provides a clear glimpse at the technology hospitals might expect to have in ten years' time and demonstrates that this type of bio–printing might be possible on a human scale someday soon. Equally exciting, being able to bio–print tissue and organs give surgeons opportunities to practice on exact replicas before actual surgeries and opens the door for more compatible organ transplants that our bodies are less likely to reject, since these bio–printed organs may ultimately be grown from each patient's own stem cells.

2.1.4.5 Healthcare

In addition to bio–printing organs and tissues, traditional 3D printing with plastics, rubbers, and metals plays a valuable role in healthcare. Like the fashion industry, 3D printers can be used to produce personalized and precision–made custom parts based on a patient's body size, age, sex, and lifestyle. This means that amputees can receive custom, comfortable, form–fitted prosthetics, dental patients can receive custom 3D printed crowns from 3D scans of their teeth, and patients can receive custom–fitted metal bone implants. Further, hospitals can produce custom machine or device parts like ventilator valves or nasal swabs, and doctors can create teaching and practice models for education, research, and training.

2.1.4.6 Housing

Major construction companies and organizations like Habitat for Humanity are betting big on 3D printing. Functionally almost no different than traditional Cartesian FDM 3D printers, there are currently giant 3D printers the size of a small plot of land that can 3D print entire houses out of unique concrete composites. In such a way, organizations can soon 3D print a house in 24 h, and a small neighborhood in a week. This type of 3D printing has profound implications for affordable housing and housing in developing nations. Most notably, it's cheaper, faster, safer, and more efficient-requiring less material with minimal waste, a smaller ecological footprint, and fewer on–site construction staff to complete.

2.1.4.7 Aerospace, Automotive

When every pound of a rocket resupply's payload counts, emailing up the 3D schematics for a wrench to the international space station is quite a bit easier than having to send different tools up with every shipment. Astronauts aboard the space station today have 3D printers, and are able to 3D print those emailed plans to build critical tools and equipment as they need them. Future explorers will be able to harness local materials to 3D print houses and equipment. In fact, the winning design from NASA's Centennial Challenges competition proposed building 3D printed habitats for deep space exploration like Mars using Martian soil. Rather than ship cement and building materials to Mars, these early Martian 3D printers will use Martian "regolith" (the layers of loose rock and soil on the Martian surface) to 3D print houses, landing pads, roads, and other structures for the next generation of space explorers.

Companies such as Ford, Porsche, Tesla, Volkswagen, Volvo, and even NASA's Perseverance Rover use 3D printed parts on their vehicles. From proof–of–concept prototypes and designs to custom–built seats, brackets, nuts and bolts, tooling equipment, fixtures, and complex components are being built directly on the assembly line floor. 3D printing opens up the door to rapidly–produce custom high–precision parts for vehicles, while also helping ease and address supply chain issues to boot.

2.2   FDM 3D Printing

In this section, we'll explore the FDM 3D printing process in more detail. Each category of 3D printing has a long list of strengths, weaknesses, applications, and unique operating steps that could fill several books, so for simplicity's sake throughout the rest of the book, we'll focus primarily on the technology you're most likely to bump into in your 3D printing journey: FDM 3D printing. To deep dive into the world of 3D printing, consider reading "3D Printing I: Introduction to Desktop 3D Printing" by Tyler Kerr.

2.2.1   How FDM 3D Print Works

Fused deposition modeling ("FDM") 3D printers are incredibly versatile machines, capable of creating everything from board game pieces to prosthetic limbs and, as we discussed, even houses. As 3D printers become more common in businesses, schools, and homes, it's important to know how to use these machines for personal or professional projects. Read on to learn about how FDM 3D printing works, common FDM 3D printer anatomy, and tips and tricks for working with typical 3D printing materials.

2.2.2   Variations in FDM 3D Printer Designs

Because they're inexpensive and simple to design, FDM printers come in a wide array of styles and shapes. There are desktop 3D printers in schools used to create educational models and larger industrial FDM machines with built–in oven enclosures that can reach incredibly high temperatures to print materials to send up into space. There are even food FDM 3D printers that can print objects out of chocolate or broccoli paste in Michelin–starred restaurants. Across these distinct variations of FDM 3D printers, the technology and design of the printer can differ.

Variations in FDM 3D printing revolve around the systems of movement for all three axes on a printer. The two most common variations in FDM printers are boxy "Cartesian" 3D printers (Fig. 2.5), which are named after the dimensional coordinate system (the X, Y, and Z–axis), and "Delta" 3D printers, which have circular build plates and an extruder suspended by three arms in a triangular configuration. More recently, there are robotic arm FDM printers and "Polar" FDM 3D printers. The most important thing you need to know is that Cartesian printers are much more common and are the most likely type of FDM printers you may encounter on your 3D printing journey. So how do we get started using a Cartesian FDM printer? Let's walk through the steps:

Fig. 2.5 A Cartesian 3D printer (left) compared to a less–common Delta 3D printer (right) Adapted from https://makerbot.com

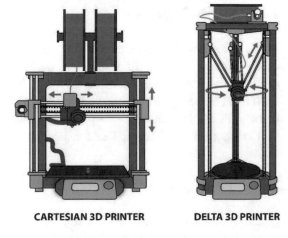

CARTESIAN 3D PRINTER   DELTA 3D PRINTER

1. Design a 3D model

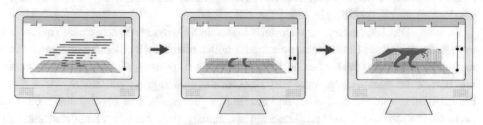

2. Slice the 3D model

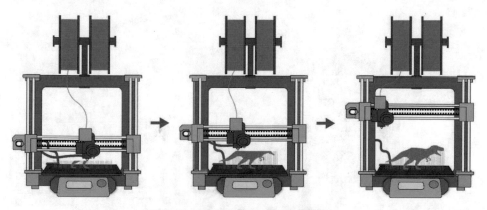

3. 3D print the sliced file

2.2.2.1 Step 1: "Slice" the File

At its core, the FDM process always begins with a digital 3D model–essentially a blueprint of the physical object. This model is sliced by 3D printing software (conveniently called "slicing software" or a "slicer") into thin, 2D layers. These layers and the instructions that you program into them (heat up to a specific temperature, print slower near the base of the model, squeeze out more plastic, add support structures) are then turned into a set of instructions in computer numerical control (CNC) machine language (a language called "G–code") for the printer to follow. You might think of this as somewhat similar to the settings you choose when using a paper printer (draft, high resolution, black and white, color). The difference here is that the 3D printing slicer stacks those 2D slices up layer–by–layer to build a 3D model. Later in this chapter, we'll dive into slicers in a bit more detail and highlight the quick and easy steps to get started with slicing software.

2.2.2.2 Step 2: Load the Material

Most FDM 3D printers can use a variety of plastics and other materials spooled up into rolls of thin filament to create prints. Common FDM printing materials that you'll likely bump into as you start 3D printing include PLA and ABS.

In most FDM 3D printers, the material is loaded into the machine and extruded out the hot nozzle. In all desktop FDM cases, an extruder with a nozzle melts filament and pushes that molten filament out of the hot nozzle. Recall that FDM 3D printing has two core components: one part of the machine (the "cold end") uses drive gears to grip the filament, which feeds the material toward the heat block and nozzle (the "hot end"). The hot end then melts and extrudes the plastic onto the bed. There are two main types of extruders that determine how the material is loaded and extruded: direct drive extruders and Bowden extruders (Fig. 2.6).

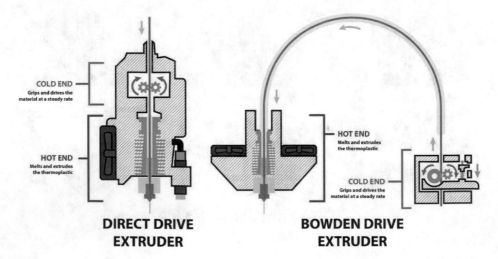

Fig. 2.6 Two most common types of extruders: a direct drive extruder (left) and a Bowden extruder (right)

- Direct drive extruders (like those on the Prusa i3 MK3S+ 3D printers detailed later in this chapter) seat the stepper motor drive gears directly next to the print head itself, so filament gets pushed directly into the hot end by the cold end. Direct drive extruders can be more reliable and easier to use, as filament is typically easier to load. They can also handle more unique types of filament. The biggest negative to direct drive extruders is that the print head must carry the weight of the stepper motor as it moves along the x–axis, which makes direct drive extruders a bit slower than Bowden extruders.
- Bowden extruders (like those on Prusa Mini 3D printers) have the drive gears on the frame of the 3D printer, typically farther away from the print head. The motor feeds the filament through a longer "Bowden tube" (usually made of PTFE plastic) to the print head. This plastic Bowden tube helps guide the filament from the cold end to the hot end. A Bowden extruder setup means that the hot end doesn't have to support the weight of the cold end, which makes the machines lighter, quieter, and typically capable of producing higher resolution, more precise prints. The downside? Driving filament a greater distance requires more powerful motors, which might drive up the cost of the 3D printer itself. Additionally, parts can be a bit stringier, with tiny spiderweb–like wisps of plastic that need to be cleaned off the final printed object.

2.2.2.3 Step 3: Start Printing

With your slicer settings selected and sent to the printer and the material loaded, it's time to start printing. Once the FDM printer receives the G–code instructions from the slicer, the hot end begins to heat up, and the printing starts. Depending on the size and complexity of the 3D printed part and the type of printer, a print might take anywhere from 30 min to 30 h to complete. For example, a simple plum–sized object might take around 2.5 h, an apple–sized object might take roughly 5–6 h, and a cantaloupe–sized object might take 12 h. All told, the details of the object and the settings you've selected in the slicer all play a significant role in the time it will take to print. To ensure everything is laying down and that material is extruding correctly, it's usually best to stick around and watch your project for a little bit to ensure there are no issues with the first layer of your print. Depending on the size of your project, this usually takes around 10–30 min. In a way, a 3D printed part is like a house: it can't stand without a sturdy, well–laid foundation. So make sure that the foundation is level and laying down well.

2.2.2.4 Step 4: Post–processing

Once your 3D printed part is finished, it might still require some touch–ups. Depending on your settings and the printer you used, 3D printed parts may not be ready to use immediately after the printing process stops. They may require some post–processing to achieve the

Soluble supports   **Supports removed**

Fig. 2.7 A 3D print with the soluble supports still attached (left) compared to a print with the soluble supports dissolved away (right)

desired level of surface finish. These steps take additional time and usually some manual effort to remove breakaway plastic supports or bed adhesion components such as brims.

We've established that FDM 3D printers work by depositing layer after layer of thermo-plastic filament to create a 3D object. Each new layer is supported by the layers under it and quickly cooled by fans attached to the hot end once extruded from the nozzle. If your model has an overhanging part that is not supported by anything below it, the printer won't be able to print across that empty gap very easily, even with hot end fans cooling the plastic in place. You will need to turn on additional 3D printing supports to ensure a successful print. Once the print is finished, you can usually break off or dissolve away these supports, leaving only the original model. The downside of supports is that if they're too close to the model, not all the support material may break off–ultimately leaving some scarring (Fig. 2.7) on the bottom of your print that you might need to sand away.

On the other hand, bed adhesion (Fig. 2.8) is the ability of a 3D printed part to stick to the build plate while printing. When 3D prints have trouble sticking to the build plate, it's usually because they're too small and haven't had enough time to settle and cool between layers, because the build plate didn't level correctly, or because the plate isn't clean. When your project does not adhere to the bed properly, you can get curled, warped, and poor–quality results. Many times, the print will simply fail. 3D printers use varying methods to ensure that objects stick to the plate while printing. Most commonly, using bed adhesion methods such as "skirts" or "brims" helps prime the nozzle with melted filament and increases the surface area of plastic touching the build plate. Any type of build plate adhesion would need to be peeled away and potentially trimmed from the model after the print is finished. We'll tackle build plate adhesion options in more detail later in the chapter.

Fig. 2.8 Three types of bed adhesion: skirts, brims, and rafts

2.2.3   Common Cartesian Printer Anatomy

2.2.3.1 A Basic Overview

Here, we'll cover the basics of common Cartesian desktop FDM printers that you're likely to bump into on your journey towards 3D printing machine mastery. Between the names of the parts and their functions, it might seem challenging to keep it all straight, but we'll keep it super simple here with a quick guide of the mechanical and electrical components found in a generic FDM 3D printer (Fig. 2.9). Later in the chapter, we'll deep dive into specific features found on Prusa 3D printers.

2.2.3.2 Axes of Movement

1. The X–axis moves left to right, carrying the extruder along linear rods or reinforced rubber timing belts.
2. The Z–axis moves the extruder up and down. These are typically driven by lead screws. Lead screws (rather than rubber belts like those that control stepper movement motion along the X– and Y–axis) allow for more fine motor control, which means more precise up and down controls, which in turn means finer resolution parts.
3. The Y–axis moves forward and backward. On Prusa printers, this moves the build plate (5) forward and backward using a timing belt. On other printers, the Y–axis may move the extruder forward and backward instead.
4. The entire hot end assembly (in Fig. 2.9: a direct drive extruder) is made up of a few different components:

   (a) The heat sink helps to dissipate the heat created by the heat block. It's separated by a small heat break, and prevents the intense 200–250 °C heat from creeping up the hot end too far, which can cause bad clogs. It's always best to turn off a printer that's sitting at temperature if you're not going to use it immediately. Otherwise, heat creep can occur, and the hot end can clog.
   (b) The heat block provides the heat to the nozzle through a heater plug and a thermistor.
   (c) The nozzle heats up and melts the filament with help from the heat block. The nozzle is screwed into the heater block. Nozzle diameters come in many sizes ranging from 0.25 mm to as wide as 1.2 mm or beyond, depending on the application. The most

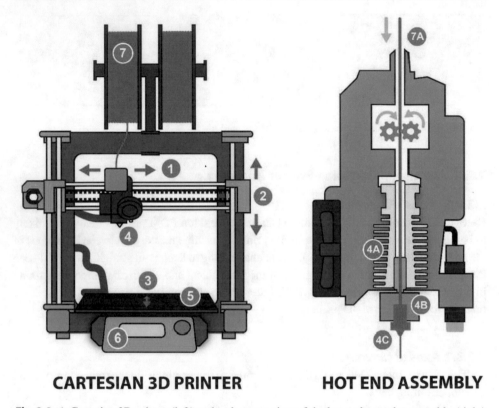

CARTESIAN 3D PRINTER   **HOT END ASSEMBLY**

Fig. 2.9 A Cartesian 3D printer (left) and a close–up view of the hot end extruder assembly (right). Figure used courtesy of Prusa (https://prusa.com)

common nozzle size is 0.4 mm, which Prusa printers typically use. Always be cautious around nozzles. Depending on the material used, they can reach upwards of 250–300 °C.

5. The build plate is where the material is extruded to form the 3D printed part. On many types of FDM printers, these are not only heated (which makes 3D printed parts stick better) but also removable and flexible (which makes taking your part off the printer a breeze).
6. The user control interface is usually an LCD screen with buttons to navigate the menu. Some printers are controlled by nearby computers while others are operated entirely machine–side.
7. The filament spool is where the filament is placed during printing. This is usually found above, behind, or to the side of the printer. Filament (7A) is fed into the extruder, heated up by the heat block (4B), and extruded out the nozzle (4C).

2.3 Affordable Desktop 3D Printers

2.3.1 Popular Brands

Today, FDM 3D printers come in all sorts of shapes and sizes. If you're reading this book and looking to learn how to integrate 3D printing into your Arduino projects–or any other projects for that matter–there are a few key considerations you should factor in so you can ensure that you're selecting the correct machine for you. What do you plan on printing? What sorts of materials or colors do you want to print with? How large will your objects be? How high resolution must your projects be?

Below, we detail a few of the more popular, reliable, and well–known 3D printing brands on the market today. Then we'll hunker down for the rest of the chapter with one of 2022s best–in–class 3D printers: the Prusa i3 MK3S+.

2.3.1.1 Prusa

Prusa printers are considered the industry standard for affordable workhorse desktop 3D printers. Machines like the Prusa i3 MK3S+ and Prusa Mini are easy to use, consistent, easy to fix, and very reliable. Better yet, company founder and inventor Josef Prusa was a core developer of RepRap and makes a point of freely sharing out Prusa 3D files in case a 3D printer breaks down. Rather than buy new parts, Prusa makes it possible to just 3D print replacement parts for free! Today, the Prusa i3 design is considered one of the most popular FDM 3D printer designs on the market, and one adopted by hundreds of thousands of manufacturers and hobbyists worldwide thanks to Josef Prusa's open–source and maker––focused approach.

2.3.1.2 Ultimaker

If you're looking for a formidable workhorse machine with lots of bells and whistles, and are prepared to pay a bit more for some serious quality upgrades, then the Ultimaker S3 and S5 printers should certainly be on your radar. For those looking for professional, dual–extrusion multi–material, and even soluble support printing capabilities, the Ultimaker printers might be top of your list. These printers are fast, sturdy, among the easiest to plug–and–play, require little maintenance, and are capable of handling almost any challenging print you might throw at them. Consider these machines if you have a bit of a larger budget and want a 3D printer that can handle tricky projects. There's a reason Ultimakers are consistently ranked as some of the best desktop 3D printers today.

2.3.1.3 Creality

Creality 3D printers such as the Creality Ender 3 V2 and Ender 3 Max are excellent budget picks for those just getting started in the world of 3D printing. Creality 3D printers can provide impressive print quality and reliability at a very reasonable price range. These machines come in a wide range of build volume sizes from small to the largest of projects and are known for being hardy and reliable. More exciting still, Creality recently came out with the first commercial 3D printing treadmill called the CR–30 that can print along the Y–axis infinitely. If you ever wanted to 3D print a full–sized sword, consider Creality.

2.3.1.4 LulzBot

LulzBot printers are well–known as workhorses in the 3D printing world. LulzBot machines, whether the compact LulzBot Taz Mini or the sizeable LulzBot Taz 6, are considered easy to use and very reliable. Despite a higher price tag, printers like the LulzBot Taz 6 and Taz Workhorse have huge build volumes and are exceptionally dependable to boot. So while hobbyist 3D printing enthusiasts might prize affordability over all else, educators and professional manufacturers looking for a solid and reliable 3D printer won't go wrong with LulzBot.

2.3.1.5 Flashforge

Entry–level Flashforge 3D printers are prized for their plug–and–play simplicity and accessibility while also known for producing reliable, high–quality parts. Machines like the Flashforge Finder are considered some of the best 3D printers for beginners, as well as young folks with little experience who are looking to dive into the world of 3D printing.

2.3.1.6 Monoprice

Monoprice printers are among the best value and beginner–friendly 3D printers that you may be able to find. Home hobbyists and educators will get a lot of use out of Monoprice machines, such as the budget–friendly, preassembled Monoprice Voxel. These machines are easy to use, produce high–quality parts, and offer a fair number of features for the price.

2.3.2   Getting Started with Prusa

2.3.2.1 Why Prusas?

There are few brands with the impeccable reputation and enthusiastic fan base that Prague––based Prusa Research enjoys. Prusa printers (Fig. 2.10) are among the best desktop 3D printers on the market, some of the most reliable, workhorse machines, and surprisingly some of the most affordable, ranging from $300 to $900. They're accessible, easy to use, simple to repair and troubleshoot, and easy to modify, thanks in part to Prusa's willingness to make all their hardware, CAD files, and software freely available online. Prusa printers are very much in line with the open–source 'maker' spirit ingrained in many aspects of makerspace culture today.

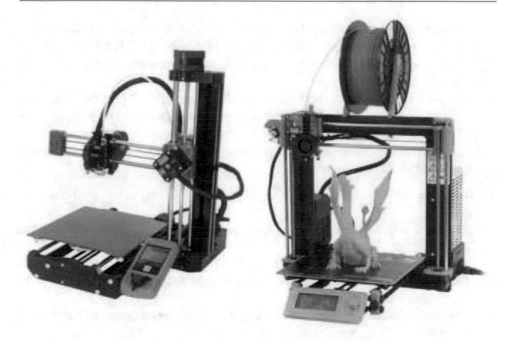

Fig. 2.10 Two popular models of Prusa 3D printers, the Prusa Mini (left) and the Prusa i3 (right). Figure used courtesy of Prusa (https://prusa.com)

Since 2017, Prusa printers such as the Prusa i3 MK3S, Prusa i3 MK3S+, and the Prusa Mini have consistently won awards and dominated the charts for "best printer" and "Editor's Choice" as some of the highest–rated 3D printers on the market. As of 2022, the Prusa i3 MK3S+ is still considered the "Best 3D Printer," according to All3DP.com, PCMag, and the industry–leading Make: Magazine. Simply put, these printers work, and they work very well.

2.3.2.2 Capabilities

The Original Prusa i3 MK3S+ 3D printer is one of the more recent designs to come out of Prusa Research in the last two years and features exceptionally high print qualities, with prints as fine as 0.05 mm (50 μm) layer heights. They have a large build volume of 9.84 × 8.3 × 8.3 inches and a direct drive extruder that can travel up to 200 mm/s, making them speedy 3D printers to boot. Prusa i3 MK3S+ machines are capable of automatic bed leveling, meaning that there's no need to manually level the bed by hand. MK3S+ build plates are also heated, which does wonders to improve bed adhesion and guarantee successful prints. Better still, the build plates are magnetic, removable, and flexible, making removing 3D printed parts very easy. Just gently flex the sheet and your print pops right off. Prusas use the more common, affordable 1.75 mm diameter filament, meaning that you can source inexpensive filament from reputable vendors fairly easily. And the Prusa i3 MK3S+ supports an impressive array of materials: PLA, ABS, PET, HIPS, TPU, PC, PP, Nylon, wood composites, metal composites, ASA, carbon–fiber enhanced filaments, and many more.

Here, we'll take a deep dive into Prusa i3 MK3S+ anatomy (Fig. 2.11). Even if you have your eye on a different FDM printer, most FDM printers fundamentally operate the same way. By learning about the Prusa i3 MK3S+, you can still learn critical components that will help you in your 3D printing journey no matter what FDM machine you use. Let's check out the Prusa i3 MK3S+ in more detail.

General Anatomy

1. The extruder moves along the X–axis, left to right, using one rubber timing belt and two linear rods.
2. The extruder moves along the Z–axis, up and down, using two threaded rods and two linear guiding rods, one pair per side. Threaded rods, rather than belts, allow for better and more precise movement.
3. The heated flexible build plate moves along the Y–axis, forwards and backward, using a belt and two linear rods.
4. More recent Prusa printers have heated, flexible, removable magnetic build plates. These make it super easy to remove finished prints: just remove and gently bend the build plate and the print pops right off.
5. The hot end assembly is the part of a 3D printer that melts the filament and helps keep the machine at a consistent and accurate temperature to ensure successful prints. It's made up of several key components. Specific to Prusa printers, the hot end is made up of a direct drive extruder with two hobbed bolts (instead of a hobbed bolt and an idler bearing), a heat sink, a heat block, a brass 0.4 mm diameter nozzle, and two fans. One fan pointed at the heat sink to help it cool, another aimed at the extruded plastic.
6. One stepper motor controls the belt on the X–axis.
7. Two stepper motors control the up and down motion of the extruder along the Z–axis.
8. Threaded rods instead of belts along the Z–axis help to provide more precise up and down motion.
9. The On/Off switch is on the back of the power supply.
10. An LCD knob is used to rotate clockwise or counterclockwise and navigate the menu.
11. A reset button under the LCD knob allows you to click to cancel or turn off the heating options.
12. Navigate the LCD panel menu to load and unload the filament and load G–code projects from SD cards.
13. Prusa i3 MK3S+ machines use SD cards loaded into the LCD panel's left side.
14. Filament spools are placed on the filament holder.

Fig. 2.11 Basic components of a Prusa i3 MK3S 3D printer. Figure used courtesy of Prusa (https://prusa.com)

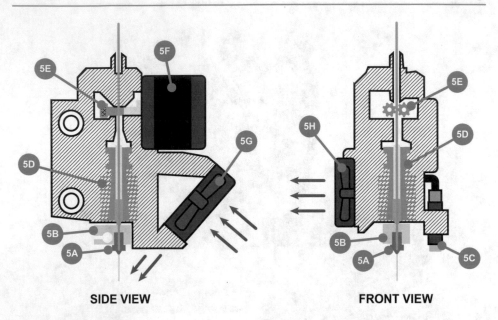

SIDE VIEW   **FRONT VIEW**

Fig. 2.12 Side view (left) and front view (right) of a Prusa i3 MK3S+ hot end. Figure used courtesy of Prusa (https://prusa.com)

15. Prusas use 1.75 mm diameter filament. It's always a good habit to unload and store filament in airtight bins. Moisture in humid areas can seep into filament and cause prints to fail.

Prusa hot end assembly Recall that the hot end assembly is the part of a 3D printer that melts the filament and helps keep the machine at a consistent and accurate temperature to ensure successful prints. It's made up of several key components. Specific to Prusa i3 MK3S+ printers, the hot end (Fig. 2.12) is made up of a direct drive extruder with two hobbed bolts, a heat sink, a heat block, a brass 0.4 mm diameter nozzle, and two fans (one pointed at the heat sink to help it cool, one aimed at the extruded plastic).

- 5A. The standard nozzle used by default in most Prusa printers is 0.4 mm in diameter.
- 5B. Above the nozzle is the heat block which, as its name implies, provides heat to the nozzle via a thermistor.
- 5C. The PINDA probe, short for "Prusa INDuction Autoleveling sensor," is a vital component to help calibrate the Prusa printers. It allows the printer to detect how far the nozzle is from the build plate.
- 5D. The heat sink helps to dissipate the high temperatures created by the heating block. If the printer is on at temperature for too long without filament moving through at a steady rate, heat creep can occur. Heat creep is the process of heat spreading irregularly

throughout the hot end, which may create clogs and ruin the hot end. Therefore, it's a good habit to ensure that the extruder is not sitting at temperature when not in use.

- 5E. The hobbed bolts have teeth to grip and pull filament down through the direct drive extruder, past the heat sink and heat block, and down towards the nozzle. Newer Prusa models have two hobbed bolts to grip filament and provide tension, rather than a hobbed bolt to grip and an idler bearing to provide tension found in older Prusa machines (and more typical in most FDM printers). If filament is grinding, the hobbed bolt is often the culprit. Perhaps there's too much pressure? Too little? Maybe there's a filament tangle that's preventing the filament from moving, so it's just grinding in place.
- 5F. The extruder stepper motor drives one of the hobbed bolts.
- 5G. A large fan aimed at the nozzle helps the cool down filament as it's extruded.
- 5H. Another fan aimed away from the heat sink helps pull heat from the hot end.

2.4   Materials

In this section we'll explore the wide variety of 3D printing materials.

2.4.1   PLA (Polylactic Acid)

PLA is the most popular 3D printer filament type due to its ease of use, affordability, and dimensional accuracy. It's one of the easiest thermoplastics to work with, and can be used on a wider variety of extruders and build plates due to its lower printing temperatures and resistance to warping and curling. Because of its popularity, PLA is available in a nearly endless abundance of colors and styles, including exotic or specialty filaments such as conductive filament, glow in the dark filament, and even filaments infused with metal or wood. Finally, PLA is a biodegradable thermoplastic usually made from corn, beets, or potatoes which is compostable in advanced commercial compost facilities. Thus, as far as 3D printing is concerned, PLA is one of the more environmentally–friendly filaments you can use.

2.4.2   ABS (Acetonitrile Butadiene Styrene)

If you have toys, play music, own home appliances, watch TV, or use a computer, you've already interacted with ABS. This plastic is one of the most popular plastics used in common injection molding today, meaning that you'll find ABS everywhere from home appliances, keyboards and computers, TVs, home appliances, toys, medical devices, pipes, and vehicles. ABS is known to be strong and heat–resistant and has great mechanical properties. The downsides of ABS are simply that it's a bit more prone to warping and curling, and a bit more fickle to dial in perfect settings compared to PLA. Because ABS is prone to warping and requires a more regulated temperature, open–air 3D printers (such as Prusa, Creality, Lulzbot, Monoprice machines) may struggle with this plastic compared to enclosed printers such as Ultimaker or Flashforge machines.

2.4.3 PETG (Glycol Modified Polyethylene Terephthalate)

Known for its flexibility, impact resistance, and as one of the most popular plastics in the world, PETG is another common type of filament you may want to use. You're likely to bump into PET and PETG with water bottles, food and medical packaging, and medical applications. Because PETG is nontoxic and FDA–approved, it's an excellent filament choice for food–related or medical projects. Worth noting: PETG can be challenging to dial in on a 3D printer and may adhere firmly to the build plate after printing. By its nature, you're also likely to have wispy 'spider web' stringing between parts of your project, which typically requires minor post–processing to remove and possibly sand down.

2.4.4 TPU (Thermoplastic Polyurethane)

If soft, rubbery, flexible 3D printing is an interest, consider using thermoplastic polyurethane (TPU). TPU is usually a composite blend of rubber and plastic, which makes this 3D printing material easier to bend, stretch, and flex. You can purchase different TPU filaments that vary in their elasticity and rigidity, with elastic properties similar to car tires through rubber bands. TPU is a great material to use for prototype drive belts, footwear, mobile tablet and phone cases, medical devices, handles and power tools, instrument panels, and even sporting goods.

2.4.5 Exotics/Specialty Filaments and Their Applications

Worth a brief mention is the wide range of rare, "exotic" filaments available. For the adventurous folks among us, there are a huge range of different experimental filaments to play with: metal–infused plastic filament that can be polished and oxidized, wood–infused filament that can be printed with a grain and sanded or stained, conductive filament, rainbow filament, marble filament, glow–in–the–dark filament, color–changing filament, shiny and reflective filament, coffee and beer filament made from food byproducts, and even soluble filament that can be used to print support structures that are later dissolved away in water.

2.5 Slicers

2.5.1 What Are Slicers?

What comes to mind when you hear the word "slicer?" Pizza, cake, and bagels might, but in the world of 3D printing, a slicer is a type of software that converts digital 3D models into printing instructions that a 3D printer reads to create an object. There are many slicers out there, but the most common slicers include Ultimaker Cura, PrusaSlicer, Simplify3D, OctoPrint, and Slic3r.

These slicers do exactly what they describe—they slice. The software takes an uploaded 3D model and cuts that 3D model into horizontal stacked layers based on the settings selected (Fig. 2.13). The slicer calculates how much material the printer will need to extrude to build the model, how thick each horizontal layer will be, and how long it will take to complete the project. This information is then packaged up into a G–code file and sent to the printer. G–code, in short, is a computer numerical control (CNC) programming language used in computer-aided manufacturing to control machines such as CNC lathes, mills, routers, and 3D printers. Slicer settings strongly influence the quality of your project, so it's important to select the best software and settings to get you the best quality 3D prints.

There are a large number of slicing softwares online and available to download, many of which are free. Even better, slicer software is surprisingly as easy as traditional 2D paper printing to use. Just load the model, select your desired settings, and hit print. At their core, slicers are simply a way to get a digital file from the computer to the 3D printer in a language that the 3D printing hardware understands.

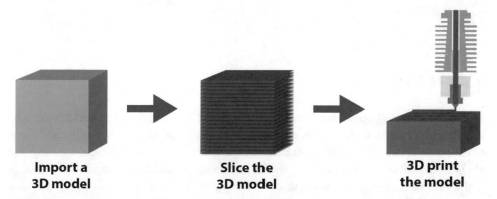

Import a 3D model → **Slice the 3D model** → **3D print the model**

Fig. 2.13 The core concept behind slicing software, where a 3D model is split into thin layers and sent over to a 3D printer

This section will cover the basic slicer settings that every beginner should know. Later, we'll cover the steps necessary to start your 3D printing project, such as loading, unloading, and storing filament, as well as navigating the Prusa menu and monitoring the initial layers of the project.

2.5.2   Where to Find 3D Models

One of the great aspects of 3D printing is that you don't necessarily have to create models all by yourself. However, if you need to create a custom tool or part for a specific prototype or design project, it certainly pays to have a basic knowledge of CAD modeling to 3D model that object from scratch. For introductory 3D printing, however, this may not be the case. For those just getting started, there are a large number of online repositories with millions of 3D files for you to access, download, and even modify. Consider visiting:

- https://thingiverse.com/
- https://grabcad.com/library
- https://www.turbosquid.com/
- https://sketchfab.com/
- https://www.cgtrader.com/
- https://www.prusaprinters.org/
- https://www.myminifactory.com/
- https://3dprint.nih.gov/
- https://www.yeggi.com/
- https://3d.si.edu/.

2.5.3   CAD Models

If you want to build custom–made models from scratch, particularly 3D models that must be a specific dimension or tolerance, computer–aided design (CAD) software is often necessary. CAD software makes it possible to build objects to the specifications you need, meaning that CAD is incredibly useful for making repairs and building custom parts or fittings. Often, as is the case with other 3D file types shared online, you may be able to use existing libraries such as GrabCAD or Onshape to find an approximate design that you can then modify. We won't dive into CAD too much in this chapter other than highlighting its use in customizing our own Arduino robot chassis later using free CAD files available online.

2.5.4   Popular Slicers and How to Get Started

For the rest of the chapter, we will use Ultimaker Cura software, a powerful and very popular free slicer (Fig. 2.14). There are a large number of slicing softwares out there, many of which are free, so you might wish to use a different program. PrusaSlicer, the slicer developed specifically around Prusa machines, is another fine alternative. We'll focus primarily on Cura since it's more applicable across a wider array of 3D printer makes and models. Whatever your slicer of choice may be, the principle settings and concepts you should understand remain the same.

Cura is one of the most popular and accessible slicers, and a favorite with beginners and advanced users alike. Cura may seem very simple at first glance, with only five or so settings to change, but hidden underneath the simple and accessible Cura exterior, however, is a treasure trove of hundreds of advanced settings to customize if you need and want to. If you're someone who likes to tinker with the 3D printer settings, you certainly can. But for those who simply want to print and don't care for all the bells and whistles, it's as easy as changing a few options, and you're ready to go. To download Cura for free, visit https://ultimaker.com/software/ultimaker-cura.

2.5.4.1 Selecting Your Printer
When you load Cura for the first time, you'll be asked to select a printer (Fig. 2.15). Later, if you want to set up an additional printer, you can navigate to Settings > Printer and add or

Fig. 2.14 Ultimaker Cura is one of the most popular free slicers available today Image used courtesy of Ultimaker (https://ultimaker.com)

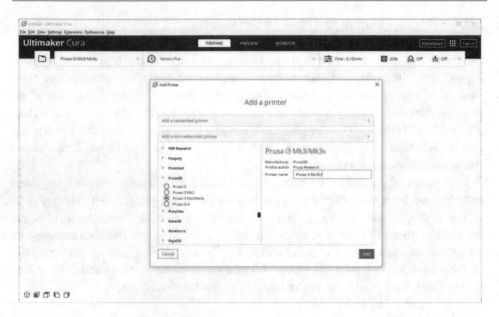

Fig. 2.15 Adding a specific 3D printer make and model in Ultimaker Cura. Screenshot from Ultimaker Cura (https://ultimaker.com)

manage printers. Because Cura is so widely compatible across a wider array of 3D printer manufacturers, the selection list for printers is long and comprehensive and will likely have your machine settings available. Most printers listed in the Cura printer selection panel already come preprogrammed with their specific features and dimensions, which saves you quite a bit of work. There's no need to program in your build volume, startup instructions, or preferred G–code language. Cura does all of that for you.

2.5.4.2 Importing a 3D File

To import a 3D model, you can simply drag and drop a 3D model onto the digital build plate. You can also select the folder icon on the left or navigate to File > Open File(s) from the top menu (Fig. 2.16). From here, you'll need to select the proper file type, as most slicers have a limited number of preferred 3D file types they can work with. STL, OBJ, or 3MF files are most typical, though increasingly CAD files are able to be directly imported. Note here that Cura works in metric, so if you've created a CAD file or 3D model in American software, you may have to do a bit of quick math to convert to millimeters.

2.5.4.3 Moving, Scaling, or Rotating the Model

When the model loads in the build area, it may be too small or too big for your project end goals. You may want to rotate the model at an angle, move it around the build plate to make

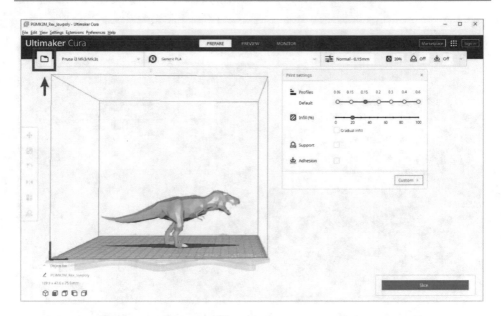

Fig. 2.16 To import a 3D model into Ultimaker Cura, click the folder icon, or navigate to File >
Open File(s). Screenshot from Ultimaker Cura (https://ultimaker.com)

room, or lay a specific planar surface flat on the build plate. Cura's lefthand toolbar allows
you to adjust your 3D model in several ways (Fig. 2.17):

- The **Move** button (T) allows you to click and drag your 3D model anywhere on the build
  plate.
- The **Scale** button (S) allows you to scale up or down the model by hand, by metric
  measurement, or by a percentage of the original file's size.
- The **Rotate** button (R) allows you to rotate the model freely, snap it to certain degrees of
  rotation, and even select specific faces of the model that you wish to lay flat against the
  build plate.

In addition to standard orientation buttons, you can click a specific 3D model with the
right mouse button in order to access additional tools. If you want to 3D print duplicates of
an object, you can click the model with the right mouse button and select *Multiply Selected*.
Cura will duplicate and automatically reposition all models. If there's enough space to print
all models, all objects on the build plate will appear in color. If there isn't enough space, the
objects that cannot fit will be shaded with gray and yellow stripes. Gray and yellow–striped
models will be ignored, and will not print.

Lastly, you might want to import a large number of files into Cura. While you can
manually position these around the build plate yourself, you don't have to. To easily and

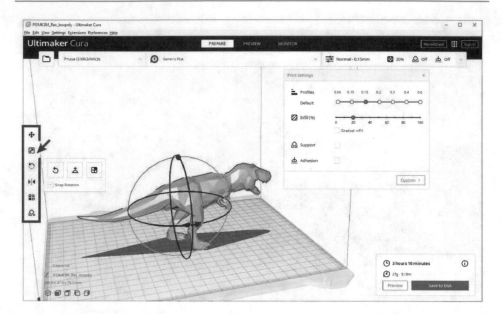

Fig. 2.17 The toolbar on the left allows you to move, scale, and rotate 3D models in Ultimaker Cura. Screenshot from Ultimaker Cura (https://ultimaker.com)

automatically arrange all models on the build plate, use the right mouse button to select the build plate itself and select *Arrange All Models* (Control + R). From here, you can make minor position changes as desired. Once you've arranged and oriented your 3D files the way you'd like them to print, it's time to select some basic settings.

2.5.5   Common Slicer Settings

2.5.5.1 Print and Quality Settings Panels

We've learned how to utilize Cura's toolbar on the left side of the window to move, arrange, and align 3D models. The menu panel along the top immediately above the digital build plate plays another critical role in preparing your 3D project to print. Three buttons (*Printer, Material, and Quality Settings*) are perhaps the most important settings you'll need to consider to get your desired print quality.

- The **Printer** dropdown menu allows you to change between 3D printers, as well as manage and add new printers.
- The **Material** dropdown menu lets you quickly select different thermoplastics such as PLA or ABS, all of which have different material properties that will affect available settings.

- The **Quality Settings** menu include layer height resolution profiles (extra fine, fine, normal, draft), infill, supports, and adhesion. There are two menu options under Quality Settings: *Recommended and Custom*. Cura will default to *Recommended* on initial startup, which is the best choice if you're just starting out learning to 3D print. Here, options are limited to only the four most crucial settings: layer height, infill, build plate adhesion, and basic support structures. Custom print options are available for advanced users who may wish to adjust hundreds of customizable settings. You can control almost every aspect of your 3D printing project using these custom settings from initial first layer fan speed through custom–designed supports. For now, we'll stick to the *Recommended* panel.

2.5.5.2 Quality Settings

The print quality setting's menu options are the most important print parameter settings for you to consider. This section will review each setting in detail and discuss why you might want to change each setting.

Profiles (layer height) The layer height of a project dictates the resolution of your print, and specifies the height of each new layer of filament extruded by the nozzle. Parts printed with smaller layers will create more detailed, higher–resolution prints with a smoother surface (Fig. 2.18). In the case of these higher–resolution prints, it may be difficult to see the individual filament layers, and may appear close in quality to a smooth injection molded part. The trade–off, however, is the time that it would take to print a part with a small layer height: since more layers make up the object, print time may double or even triple.

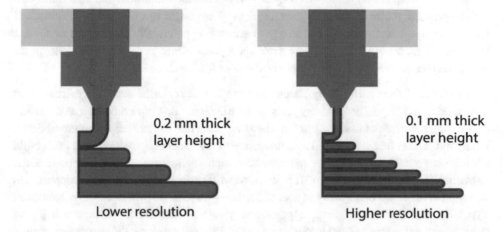

Fig. 2.18 The resolution of a 3D print is determined by the size (thickness) of each layer on top of the other. Screenshot from Ultimaker Cura (https://ultimaker.com)

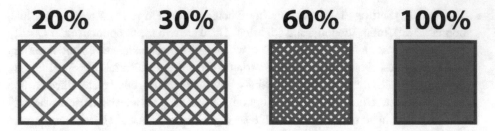

Fig. 2.19 Infill is the structural latticework inside the 3D print itself. Denser infill leads to stronger parts, but at the expense of time and material

If you wish to print something at a standard resolution, you might consider selecting a 0.15, 0.2, or even 0.3 mm resolution. These parts will have thicker layers and a rougher surface finish where individual layers may be clearly visible. Parts printed at these normal or draft resolutions will print faster than 0.1 mm or 0.06 mm resolution prints. This type of standard or low–resolution printing is fantastic for rapid prototyping projects where capturing extra fine details may not be as important. For most projects, a standard 0.15 or 0.2 mm layer height is just fine.

Infill density Infill refers to the density of the latticework of support structure inside a 3D printed object (Fig. 2.19). This infill provides structural rigidity and allows the model to print complex internal structures, roofs, and walls of the 3D model. If an object is printed with 100% infill, it will be completely solid on the inside. A part printed with 0% infill will be fully hollow and often thin and delicate. There's a trade–off: the higher the infill density, the stronger and heavier the object will be, but the more time and material it will take to print. If you're creating an item for display, we might recommend 5–20% infill. If you need something functional and sturdy, or something potentially undergoing significant stress or strain, 50–70% infill may be more appropriate. Only in specific situations do we find ourselves needing to print anything at 100% or 0% infill.

Supports Recall from earlier in the chapter that supports are external structures that surround your model and help to prop up any parts of the 3D model that have too steep an overhang or too long a horizontal bridge. Think of these supports like the latticework of crossbeams, rebar, and girders that would hold up a bridge currently under construction. You can't build a bridge out into horizontal empty space without supports holding up those new parts of the structure. The same rule applies to 3D prints. Once a 3D print is done, the plastic supports can be snapped off or dissolved away in water depending on your 3D printer and its capabilities. Typically, you'll need to toggle on supports for any object with overhanging parts or bridging components extending beyond a 45–degree angle. Without supports, the parts might droop, sag, or even fall over, which may ruin the project. Supports on Prusa printers are most often printed in the same material as the model itself. These supports are designed as "breakaway" supports – that is, they are designed to break off easily once the project is finished.

Fig. 2.20 The Y–H–T rule. Parts with bridges or overhands (H, T) typically require support, while parts that gradually build upon each other (Y) may not

How do you know whether your design needs supports? Cura will tell you in the *Prepare* view by highlighting overhangs and bridges in red on the model. Another handy trick is the common "Y–H–T" rule (Fig. 2.20). If you were to 3D print the letters Y, H, and T, each might require different support settings:

- "Y"–shaped features with gradually sloping parts may be acceptable to print without supports toggled on because there is likely sufficient material overlap from layer to layer to keep the model from failing. Any slopes steeper than around 45–60° may still need supports, however.
- "H"–shaped features with a horizontal bridging structure should have supports underneath that bridge to prevent failure or a messy, stringy final product.
- "T"–shaped features similarly would need supports under overhanging parts on both branches of the "T" to avoid failure.

Build plate adhesion Whether or not you toggle on build plate adhesion and the build plate adhesion type you choose may affect the likelihood of your print succeeding or failing. Build plate adhesion affects how well your 3D model sticks to the build plate. If parts are too small or have minimal surface area touching the build plate, there may be an increased chance for your print to become detached from the build plate. When in doubt, it doesn't hurt to have adhesion toggled on. By default, Cura will enable a "brim" for the build plate adhesion type if you toggle on this setting.

Like a brim of a hat, 3D printed **brims** are the lines around the bottom of an object that help increase the surface area of the first layer touching the build plate, and also help to keep the corners of your model from peeling or warping up (Fig. 2.21). Brims can also stabilize or link together delicate parts of an object that may be isolated from the rest of the project.

Fig. 2.21 Brims are a common type of bed adhesion, and are used if projects are having trouble adhering to the build plate

2.5.5.3 Viewing Your Selected Settings

Up until now, we've largely been working in Cura's *Prepare* view. This allows us to change many settings on the fly without having to inspect the model after every change. With your layer height, infill, support, and adhesion settings now selected, it's time to take a look at Cura's other view panels. Select the *Slice* button in the bottom right corner to commit your settings. Next, we'll navigate to *Preview* at the top to review our model with the settings implemented. Worth noting, in Cura, there are three ways to view the model, each of which has a different utility that you might find useful. The default view when importing files, *Prepare* gives you a good idea of how the digital model appears, and allows you to change important project settings. In contrast, the *Preview* tab we've just selected provides two view types: Layer view and *X–ray* view.

Within the *Preview* tab, *X–ray* view provides a translucent view of your 3D model, which may be useful if you need to inspect a 3D file for missing or glitch geometry. We rarely use this view type, as there are better programs out there to inspect and repair files. Layer view, however, is one of the more important Cura windows that you should utilize.

Once you've selected settings and are ready to print, you should always hop into *Preview* > *View Type*: Layer view and inspect your model to ensure that it will print correctly. Changing the color scheme to "Line Type" helps visualize support and brim structures. A slider on the right side of the UI will allow you to scroll up and down layers of the model (Fig. 2.22). As you get more comfortable in Cura, *Layer view* can be critical to ensure that your model is oriented on the build plate correctly. It's a very good habit to check the Layer view and scroll down to inspect the first few layers of your project before starting to print. Remember, 3D prints need a good foundation. If insufficient model or support material is touching the build plate, the print may not succeed.

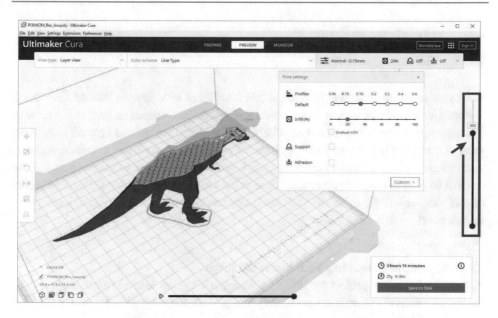

Fig. 2.22 Use Preview mode's Layer view to scroll up and down and assess your sliced model. This is a good method to ensure the success of a project. Screenshot from Ultimaker Cura (https://ultimaker.com)

2.5.5.4 Quick Summary: Slicer Tips and Tricks

– Cura has hundreds of different settings and a plethora of profiles available for dozens of different printer brands. Because of this, it may feel overwhelming when looking at the complete list of settings. Cura defaults to *Recommended* settings for users, and for the most part, this is the only list of settings you should need when you're first learning. Custom settings can come later once you're comfortable with the basics.

– Always get into the habit of double–checking your settings. As we'll discuss, once your model has been sliced, be sure to look at it in *Layer View* to ensure nothing looks amiss. It's good to double–check that your project is sitting flat on the build plate and that nothing is floating in midair. This is one of the most common reasons for failed prints.

– Remember to save your model as G–code for the printer, as well as a project file to edit later if need be. This will allow you to fix any errors that present themselves during printing without reconstructing the project settings from scratch.

– When troubleshooting, like any good engineer or scientist, it's good practice to change only one setting at a time and observe the results. This is the best way to track how each incremental change affects your print. Don't change too many settings at once.

2.6   Preparing to Print

2.6.1   General Overview

After you've sliced your file, the Slice button is replaced by a Save to Disk or Save to File button. Use this button to save your project as G–code instructions to send to the printers, and also ensure that you've saved an editable project file (File > Save Project) in case you need to make changes. Once you have your file prepared in your slicer of choice, and once you've saved your project and exported your G–code in the proper format onto an SD card or USB to plug into the 3D printer, it's time to get the printer ready to go. The final section of this chapter will give you the tools necessary to print your project. You're in the home stretch now!

2.6.1.1 Double–Checking Your Slicer Settings

As you prepare your project for the 3D printers, it's a good idea to run through a quick mental checklist to ensure that your slicer settings work for the goals you wish to achieve:

– Does my project need supports? If so, have I toggled supports on?
– Does my project have small features touching the build plate? Is there minimal surface area touching the build plate? If so, have I toggled on "Bed Adhesion?"
– Have I sliced and viewed my project in the Preview tab? Do I feel confident the first layers are oriented correctly and touching the build plate?
– Is the project set to the desired resolution and infill percentage?
– Is the project saved as a G–code file for the 3D printers in addition to a project file in case I need to make edits?

2.6.1.2 Navigating the Menu

Once you've saved your files, it's time to load filament, find your saved file, and start your project. To do so, simply plug in the SD card on the left of the LCD menu of the Prusa. It may take a second to load the projects.

Controlling the LCD screen and navigating the menu is done by a single element: a rotational LCD knob that can be pressed to confirm the selection (Fig. 2.23). Press the LCD knob to enter the main menu.

When the printer is active, the main LCD panel provides the following information:

1. Nozzle temperature (actual/desired temperature)
2. Build plate temperature (actual/desired temperature)
3. Progress of printing in % – shown only during the printing
4. Status bar (Prusa i3 MK3S+ ready./Heating/file_name.gcode, etc.)

Fig. 2.23 The Prusa i3 MK3S
LCD panel. Image adapted
from https://prusa3d.com

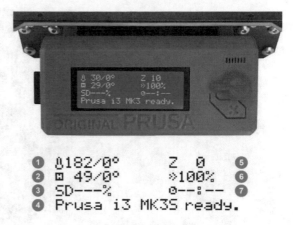

5. Z–axis position
6. Printing speed
7. Remaining time estimation

It's a good habit to always double–check your settings to ensure that the nozzle and build plate temperatures are accurate. For now, you'll want to navigate to Preheat and select your desired material. It may take about ten minutes to heat up and should beep when ready. While the printer is heating up, you can pick out the material you wish to use.

2.6.1.3 Loading Filament

Filament is mounted on a sturdy spool holder above the printer and loaded up and over the spool, from back to front. There are several diameters of filament used by different 3D printers. For example, Prusa printers use 1.75 mm diameter filament; other printers such as Ultimaker and Lulzbot machines use 2.85 mm diameter filament. Make sure to purchase the filament diameter that works for your specific machine.

If you live in a humid area, once filament has been opened, it's a good idea to store your spools in dry or atmospherically–controlled boxes. Many FDM materials can take on atmospheric moisture, which causes them to become brittle or difficult to print. If your filament is making lots of snap, crackle, and popping noises like a bowl of Rice Crispies, it has likely taken on too much moisture and will need to be dried out to guarantee an acceptable print quality.

Once you've selected a filament to use, rest the filament spool on the top of the filament holder. We'll load it once the printer heats up. Always make sure you're either holding the filament spool tightly or that the filament is secured through a series of holes or notches located on the filament spool itself. Only take the filament out of the secure filament spool holes when the printer is heated up and you're ready to load it. Accidentally letting filament unspool or unwind can cause tangles, jams, and print failures.

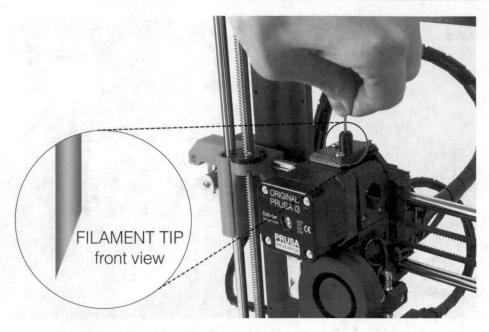

Fig. 2.24 When loading material, it's a good habit to clip the tip of your filament at an angle to prevent clogs. Image from the Prusa i3 MK3S manual Image used courtesy of https://prusa3d.com

It's also a good habit to clip the end of the filament at an angle, both before you start if it's not already clipped and after your finish and unload your filament again. While the Prusa heats up, inspect your filament. Make sure there is no blob or blob at the tip of the filament. Use clippers to remove the blob and create a nice, angled point if there is (Fig. 2.24).

Once the machine is at temperature, select the Load Filament option from the main menu, insert the filament into the hole at the top of the extruder, and push down with light but firm pressure. If the printer is actively moving the hobbed bolts to load filament, you should hear a faint noise and also feel the hobbed bolts gently grip and pull the filament. Once you feel the machine grip the filament, release your grasp and wait until filament extrudes from the nozzle in the correct color. The printer may pause after a period and ask if the filament is the correct color. Continue extruding filament until the proper color is coming out. Discard or recycle the ejected material.

Another important concept to note here is temperature. Typically, it's good practice only to have the printer sit at material melting temperatures when changing filament or actively printing. Recall from our discussion of the concept of heat creep that it's not a good idea to have the printer sitting idle at temperature for too long. In all cases, filament should be moving through the hot end at a steady rate. If the printer is stopped and not printing (or not getting ready to print), the machine should not be at temperature. If it is, it can cause heat creep and result in a jammed extruder.

2.6.1.4 Starting Your Print

Once your filament is loaded, navigate to your project from the LCD main menu by selecting Print from SD and locating the file. Before printing, be sure to inspect the printer nozzle and build plate and remove any debris from the build plate, like old primer lines (which we explain below). These can be thrown out or recycled.

Then, the exciting part: simply click your project to begin 3D printing. The printer may take a few minutes to heat up. Once it does, the printer will perform a series of nine short probes around the build plate using a special sensor. This PINDA probe knows that it should be a set distance away from the build plate at each of these nine points and can detect if the build plate is higher or lower than expected, which ultimately determines if the build plate is level enough to print (Fig. 2.25).

Once the probe has finished checking the build plate, the 3D printer will begin to print. A primer line is usually extruded at the front of the build plate before printing begins. This ensures that the nozzle has filament of the correct color and that the printer is laying down filament correctly. Primer lines, blobs, and bed adhesion skirts also exist to ensure that filament is at the very tip of the nozzle, ready to extrude. Though they're very clever machines, 3D printers aren't quite clever enough to tell if they have thermoplastic sitting at

Fig. 2.25 The PINDA probe, used to detect if the build plate is level. Many modern printers have similar bed leveling technologies. Image from the Prusa i3 MK3S manual Image used courtesy of https://prusa3d.com

the very edge of the nozzle or if some of the filament has seeped or oozed out over a short period of time, leaving an air gap inside the nozzle reservoir. If you were to print with an air bubble in the nozzle, the printer would simply move as if it were printing normally, leaving a gap or hole in your model and often causing the print to fail. The easiest way to fix this? Simply extrude some filament in a primer line or blob or skirt right before the print itself to ensure that the printer is primed and ready to go. Most 3D printers automate this process, but you can add or remove primer lines and blobs in Custom settings of Cura.

As with your slicer settings, run through a quick mental checklist to ensure that you've done everything correctly before hitting Print.

- Is the build plate clear? Is the previous primer line is removed and thrown out or recycled?
- Is the correct filament loaded? Is the color extruding correctly?
- Is the nozzle clean? It's important to ensure that there is no residual gunk stuck to it or the surrounding hot end. If there is, you should cancel your print by hitting the "X" button below the LCD knob. Then, preheat the printer for the material you're using. When the printer is at temperature, use a brass brush to very carefully and gently brush the nozzle gunk off. Avoid touching the brass brush to the fragile thermistor wires, which are usually red and white and coming out of the heat block.

It's also a good rule of thumb to supervise the first full ten minutes of any print. Why? To make sure everything is laying down and that material is extruding properly. Remember, a 3D printed part is like a house: it can't stand without a sturdy, well–laid foundation. Then, if everything looks like it's laying down well, you should be in the clear. Periodically check your project to ensure that the project is printing as intended. Congrats! You've just successfully started your first 3D print. Read on to learn what to do once your print has finished.

2.6.1.5 Unloading and Storing Filament

Since it's best to keep unused filament in dry storage, you should unload filament once the machine is finished printing. First, select the Unload Filament option from the menu. If the machine has cooled down significantly beforehand, you may need to preheat it again by selecting your material. Once ready to unload filament, the machine will beep and prompt you to click the LCD knob. It will then rapidly reverse the material out of the top of the extruder. Make sure to hold the filament and remove it from the extruder, grasping it firmly to ensure the filament doesn't accidentally unspool. If there is a blob or bulb at the end of the filament from where it began to melt while printing, use wire cutters to remove the blob and trim the end to an angle. Then firmly secure the end of the filament to the spool through the holes on the sides of the spool to prevent it from unwinding before placing it back in your storage box.

Most 3D printers like Prusas are perfectly fine to sit on and idle without using too much power. However, it is important to make sure that your machine is not sitting at temperature for long periods of time. You can press the "X" button under the LCD knob to reset the printer back to its ready state, which also resets and turns off the heating elements. To fully turn off the Prusa, locate the power switch at the back of the machine by the power transformer and switch it to the OFF position.

Once the machine has cooled down, don't forget to clean up any debris from on or around the build plate. This includes things like discarded supports or primer lines. If the hot end needs to be cleaned, it must be cleaned at temperature with a wire brush. Try your best to avoid touching the build plate with your bare hands, as the oils within your hands may affect the adherence of prints to the build plate.

If you need to cancel your print or if you suspect the machine is malfunctioning in the middle of a project, quickly cancel your print by hitting "X" or turning off the machine and then remove your incomplete project. Take some time to service the machine or clean specific components before you restart your project. Simple malfunctions might include scary noises such as the 3D printer stepper motors grinding on an axis, the nozzle coming into direct contact with the build plate, the 3D printer making repeated beeps or unexpected electronic noises, or filament failing to extrude correctly.

2.6.2   Best Practices

- Every material has different properties when printing. Be sure to consult the manufacturer's recommendations for printing parameters like temperatures and speeds if you are unsure how to use the material. Always work in a well–ventilated area, even if the material you are printing with does not create fumes.
- When removing supports from FDM 3D prints, excess material can fly off in tiny shards. It's good practice to wear safety glasses to prevent eye damage.
- FDM 3D printers must have filament loaded for the entirety of the print, or else the project will fail. Therefore, always make sure you have enough filament for the whole project.
- Many printers are limited by build volume size. If you wish to print objects larger than the build dimensions, you may need to split your project into pieces first.
- Many FDM 3D printers have safety features such as emergency stops and power loss protocols built into them. However, this does not mean that they can be left unattended for large amounts of time. Make a habit of checking on your active projects periodically.
- Importantly, you should always supervise the first full ten minutes of any print to ensure the project has a strong foundation.

2.7   Application: 3D Printed Arduino Robot

Ready to tie this all into Arduino? Let's 3D print and build the popular, open–source "Otto" Arduino robot. Using your knowledge of 3D printing, you'll be able to create and even customize the outer chassis of the robot, and eventually even use your smartphone to make Otto dance, walk, sing, and navigate around obstacles. You can buy Otto Maker kits online via https://www.ottodiy.com/store/products/otto-diy-starter and customize your own CAD designs for Otto via https://www.ottodiy.com/design (Fig. 2.26).

2.7.1   3D Printing the Otto robot.stl Files

Otto is available in two kits: a pre–made Builder kit with all 3D files pre–printed and a Maker kit, which only provides the electronic components. The Maker kit is for those readers who have access to a 3D printer and want to 3D print the parts themselves, which we'll want to do here. You can download the Maker kit 3D model files for free by visiting https://www.ottodiy.com/academy.

To start, download all the files for "Otto Starter" and load them into the Cura slicer (Fig. 2.27). For this example, you'll want to import the "new PCB" model files. If you've purchased a SparkFun Otto model, use those 3D files instead. Can you use what you've learned about 3D printing to address the following challenges?

– Otto's 3D file repository only comes with one leg .STL file. This is on purpose. Do you remember how to duplicate objects in Cura?
– There are many small parts to the Otto assembly. How can you rearrange the entire Cura build plate in one quick step to optimize the build plate for printing?
– When imported into Cura, Otto's head is oriented upside down. To save lots of time and support material, it may be best to flip the head 3D model upside down. How can you select a specific face and align it to the build plate?

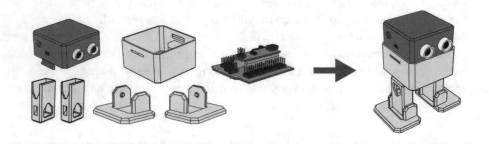

Fig. 2.26 3D printed Arduino robot

Fig. 2.27 Once you've downloaded and imported the Otto 3D files, your build plate should look something like this. Screenshot from Ultimaker Cura (https://ultimaker.com)

Since we're printing the parts ourselves, it's up to you to choose the parameters you'd like to 3D print with. Otto is well–suited to 3D printing and should 3D print very easily if you've been following this chapter closely. We recommended printing Otto using a Cartesian FDM 3D printer using the following settings:

– Material: PLA
– Resolution (Layer Height) Profile: 0.20 mm or 0.15 mm
– Infill: 20%
– Supports: None
– Bed Adhesion: None (which defaults to a skirt)

Once you've selected your settings, make sure to slice your project and head over to "Preview" mode to view your model and inspect the settings you've selected. Then, use the slider on the right side of the UI to scroll down to the base layers of your project and ensure that everything is laying flat, sufficiently contacting the build plate, and that your infill and support settings are accurate (Fig. 2.28). In our example, Otto should take around ten hours to 3D print and use about 96 g of material. Since filament spools are usually around 1000 grams of material in total and typically cost between $20 and $30 per spool, we'll be able to 3D print plenty of inexpensive replacement or custom parts if we want! Pop quiz: what

Fig. 2.28 Always make sure to check Preview mode and scroll down to the bottom of your sliced file in Layer view before pressing 'Print.' Can you recall why? Screenshot from Ultimaker Cura (https://ultimaker.com)

happens if you start to change settings in Cura? What settings might you change to shave off an hour or two and make the Otto project 3D print faster?

2.7.2 Assembling and Coding Otto

Otto is programmed to operate out of the box, once you've finished assembling the 3D printed parts. For up–to–date assembly instructions, visit https://www.ottodiy.com/academy. Once your 3D printed Otto is assembled, access the Otto Arduino IDE guide to begin coding. You'll need to download the Otto DIY libraries by visiting https://github.com/OttoDIY/OttoDIYLib/.

Next, open the Arduino IDE and navigate to Sketch > Include Library > Add .ZIP Library. Find the location where you saved the Otto DIY libraries and open the file. You'll receive a confirmation that the library has been installed. To double–check, you can navigate to Sketch > Include Library menu. You should now see the "Otto DIYLib" library at the bottom of the Include Library menu. You can also search for and install the Otto libraries via Sketch > Manage Libraries. If you're curious, you can poke around some of the base libraries hidden within the primary Otto.h library:

– Otto.h and Otto.cpp contain all the primary functions.
– Otto_gestures houses all the gestures functions.
– Otto_mouths houses all the mouth functions.
– Otto_sounds houses all the sound functions.
– Otto_matrix houses all the matrix functions.

To add the primary Otto library to your Arduino project, enter:

```
#include <Otto.h>
Otto Otto;
```

You'll also need to define each of the pins on your Arduino to correspond to legs, feet, and buzzers on your Otto robot. Declaring these pins tells the Arduino IDE software to define pin 3 as "LeftLeg" for the left leg servo motor, for example.

```
#define LeftLeg 2    //left leg pin
#define RightLeg 3   //right leg pin
#define LeftFoot 4   //left foot pin
#define RightFoot 5 //right foot pin
#define Buzzer 13    //buzzer pin
```

Next, we'll need to write some code for Otto's startup procedure under void setup. Remember that "void setup()" is the code we want to run one time as soon as the Arduino board gets power and the program starts running but then stops. This is where we'll wake up Otto and set the robot to its home state. "Otto.init" initializes the robot by syncing the leg and feet pins, as well as a few of the sensors. "Otto.home" brings Otto to its home, neutral position: legs forward, feet oriented side–to–side.

```
void setup()
{
Otto.init(LeftLeg, RightLeg, LeftFoot, RightFoot, true, Buzzer);
Otto.home();
}
```

Now the fun part. Otto has lots of customizable functions that you can change. For the sake of space, we won't detail them all here. Instead, we'll wrap up this chapter with a focus on a small subset of Otto commands that should allow you to see exactly how to start customizing code for your own personal Otto.

To import the example code, start a new project and navigate to File > Examples > OttoDIYLib > Otto_allmoves. There are tons of ways to customize your Otto, but we'll focus on movement. The servos and sensors in your assembled Otto work in tandem to let Otto shuffle forward and backward, side to side, jump, dance, sing, detect walls, and even moonwalk. You can change the values inside the parentheses for commands such as

"Otto.walk(steps, T, dir)" to make Otto walk faster, slower, change directions, and even change the size or angle of the movement. In our Otto.walk example, parameters include the number of steps in the programmed time period, time (in milliseconds), and direction (1 for forward or left depending on the command, −1 for backward or right). For example, if we were to type "Otto.walk(5, 2000, −1)," our robot would walk backward five steps over a two–second (2000 ms) period. A higher time value means a slower movement. Try values between 500 and 3000 ms. Next, try to edit some of the following. What happens to your Otto robot? Can you predict how the Otto robot might behave just by reading the code?

- Otto.walk(10, 1000,1);
- Otto.turn(5, 500,−1);. Recall here that for turning, bending, and side–to–side shaking movement, 1 and −1 represent left and right.
- Otto.bend(1, 2000, 1);
- Otto.shakeLeg(5, 3000,−1);
- Otto.jump(2, 500); For this command, there is no direction of movement. Instead, Otto will simply flex its legs to stretch upward.
- Otto.moonwalker(3, 1000, 25, 1); Note that there's one additional parameter here, between time (1000 ms) and direction of moonwalk (left, or 1). This represents a new parameter for all dance moves, "h" for the height or size of the movement. For the moonwalk, try changing the h value between 15 and 40. What happens?

A complete list of commands and their parameters can be found alongside the Otto library by visiting https://github.com/OttoDIY/OttoDIYLib/.

2.8   Summary

The goal of this chapter was to provide you with a robust background in 3D printing, with the hope of empowering you to 3D print your very own customizable Arduino robot. We first started with an overview of 3D printing technologies and then hunkered down with the most common and popular Cartesian FDM 3D printers. We reviewed popular 3D printing brands on the market today and covered the more common materials that you may be most likely to use as you first get started. Next, we learned how to navigate and change key settings in 3D printing slicer software to get our 3D models ready to print. Then, we learned exactly how to print those files. Finally, we wrapped up with an exciting application: using what you learned about 3D printing and introductory Arduino code, you should have a firm handle on how 3D printing works and how you can 3D print, customize, and code your very own Otto robot.

2.9   Problems

1. Construct a table to summarize common categories of 3D printing and their characteristic features.
2. Select an industry and describe how 3D printing may be used to enhance the industry.
3. Can you name the key components of a typical Cartesian 3D printer?
4. Summarize the steps in the 3D FDM printing process.
5. Construct a table to summarize the different types of 3D print material.
6. Describe the Y–H–T rule. What is its significance in 3D printing?
7. Can you list some of the best practices to consider while preparing to 3D print?

Additional Resources

1. Advanced Otto instructions, https://wikifactory.com/+OttoDIY/otto-diy
2. Arduino Otto instruction guides, https://create.arduino.cc/projecthub/
3. Otto DIY robot 3D files and instructions, https://www.ottodiy.com/
4. Prusa 3D printers, https://www.prusa3d.com/
5. PrusaSlicer slicer software, https://www.prusa3d.com/page/prusaslicer_424/
6. Ultimaker Cura slicer software, https://ultimaker.com/software/ultimaker-cura

Robotic Concepts and Sensors

3

Objectives: After reading this chapter, the reader should be able to do the following:

- List the important parameters a robot must sense;
- Explain how the Global Positioning Systems (GPS) may be employed to localize a robot's position outdoors;
- Describe the limitations of GPS localization;
- Describe different methods of steering a robot;
- Describe how pulse width modulation techniques are used to steer a robot;
- List different methods used by robots to sense obstacles in the environment;
- Describe techniques used to monitor and track robot motion;
- Describe sensors employed for a robot to sense its environment;
- Compare and contrast open and closed loop control; and
- Compare and contrast autonomous versus remote control.

3.1 Overview

In this chapter we investigate common concepts for robot applications. In a given application, all concepts may or may not be employed. The concepts are equally applicable from small, simple robots to complex systems. We discuss concepts related to robot location, steering, vision and obstacle avoidance, odometry, status monitoring, control, and autonomous versus remote control.

From a robot's point of view, the concepts in this chapter provide answers to questions such as:

© The Author(s), under exclusive license to Springer Nature Switzerland AG 2022   101
T. Kerr and S. Barrett, *Arduino IV: DIY Robots*, Synthesis Lectures on Digital Circuits
& Systems, https://doi.org/10.1007/978-3-031-11209-6_3

- Where am I on the Earth's surface?
- Are there obstacles nearby I need to avoid?
- How will I move about?
- What is my orientation and movement within a local X, Y, Z coordinate system?
- What direction am I heading relative to a compass setting?
- What are the characteristics of the environment around me?

3.2   GPS: Robot Localization on the Earth's Surface

In advanced robotic systems, it is useful to localize a robot's position on the Earth's surface. The Navstar Global Positioning System or GPS provides this feature. The GPS system consists of 24 satellites in medium earth orbit at 20,200 km above the Earth's surface as shown in Fig. 3.1. The GPS system provides positioning, navigation, and timing (PNT) information anywhere on the Earth's surface. The localization accuracy is approximately 4.9 m. To achieve this accuracy, a GPS unit typically receives signals from four different satellites. To receive the satellite signal the GPS receiver's antenna must be outdoors (www. gps.gov).

Adafruit provides an Ultimate GPS Logger Shield (#1272) as shown in Fig. 3.1b. The shield may be equipped with an external active GPS antenna (#960) via an SMA to FL RF adapter cable (#851). The shield provides a GPS positional fix ten times per second. GPS information is provided in the form of a variety of standard format NMEA messages. The NMEA $GPRMC message format, known as the Recommended Minimum Navigation Information (RMC) message, is shown in Fig. 3.2. The $GPRMC message provides the answer to a number of questions regarding time, location, speed, course, date, and deviation from true North [Ada GPS] [2].

A closer look at some of the definitions from the $GPRMC message are in order. A location on the Earth's surface may be specified using the latitude, longitude coordinate system. The latitude indicates how far a location on the Earth is above (or below) the equator. The latitude is specified in the form of degrees, minute, and seconds (DMS). Similarly, the longitude specifies how far away a location on the Earth is from the reference line of the prime meridian. The prime meridian (0° longitude) passes through Greenwich, London. The prime meridian also serves as the reference point for Greenwich Mean Time (GMT) as shown in Fig. 3.3. The time standard was updated in the early 1970s and is now referred to as Universal Coordinated Time (UTC) [www.greenwichmeantime.com].

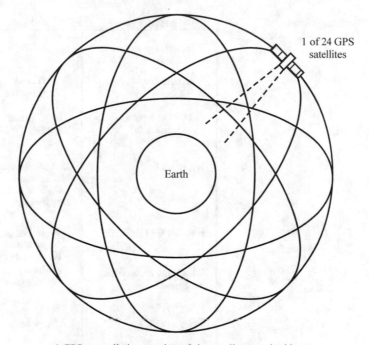

1 of 24 GPS
satellites

Earth

a) GPS constellation consists of six equally spaced orbits
with four satellites per orbit. Typically a user is in view
of four satellites at any point on Earth [www.gps.gov].

b) (left) Adafruit Ultimate GPS Logger Shield (#1272), (right) external
active GPS antenna (#960) connected to shield via SMA to uFL
RF adapter cable (#851). [Images courtesy of www.adafruit.com].

Fig. 3.1 NAVSTAR Global Positioning System (www.gps.gov) [10]

3.2.1   GPS Logger Shield

Adafruit provides excellent documentation and support for the GPS Logger Shield. A thor-ough review of their document "Adafruit Ultimate GPS Logger Shield" by Lady Ada (Ada 2020) is highly recommended. It provides a thorough step–by–step introduction to the GPS

Fig. 3.2 GPS message format
(GlobalTop) [8]

Question?	Name
What message was sent?	$GPRMC
What time is it?	UTC/GMT time hhmmss.sss
Did GPS unit get a good location fix?	Status A= valid, V = not valid
Where am I?	Latitude ddmm.mmmm N/S indicator Longitude dddmm.mmmm E/W indicator
How fast am I moving?	speed over ground knots (knot = 1.15 MPH)
What is my course over the ground?	course over ground degrees
Today's date?	ddmmyy
How close to true North?	degrees, E/W

Logger Shield. Provided here is a brief overview of steps to obtain a GPS fix and obtain data from the shield.

- Set the "Direct/Soft Serial" switch on the GPS Logger Shield to "Direct."
- Upload a blank sketch to the Arduino UNO R3.

```
//*********************************************
void setup() {}
void loop() {}
//*********************************************
```

- Set the "Direct/Soft Serial" switch on the GPS Logger Shield to "Soft Serial."
- Open the Serial Monitor and note the NMEA sentences provided by the GPS Logger Shield.
- There are a number of interesting sketches to test the GPS shield from the Adafruit GPS Library. The GPS_SoftwareSerial_EchoTest was uploaded and the resulting output is shown in Fig. 3.4.

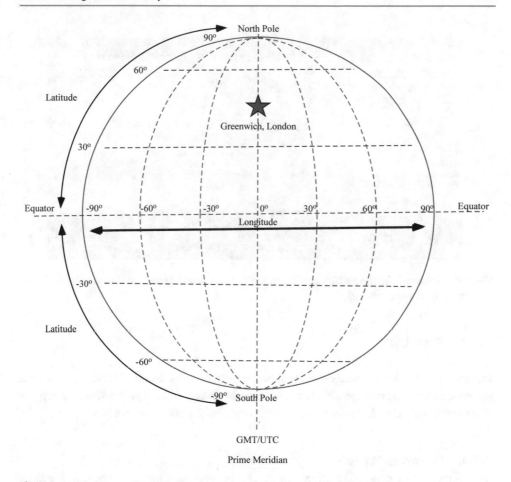

Fig. 3.3 Lat Long coordinate system (www.greenwichmeantime.com) [11]

3.3   Steering and Odometry

In this section we discuss the related concepts of robot steering and odometry. We explore the concepts of robot steering first. Based on the number of powered wheels, we investigate robot features required to render a turn. We then shift focus to the related concept of odometry. Odometry tracks distance traveled per wheel. Tracking this information allows repeatable robot turns, measurement of distance traveled, and robot velocity.

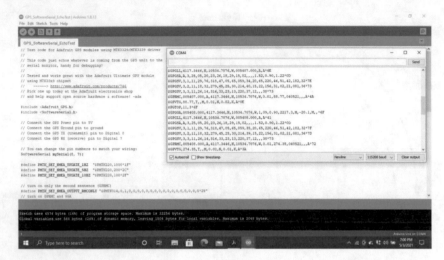

Fig. 3.4 GPS test. Note the NMEA sentences received on the serial monitor. Can you determine the locational fix of this example?

3.3.1 Steering

Figure 3.5 illustrates the fundamental robot steering concepts. Robot steering is dependent upon the number of powered wheels and whether the wheels are equipped with unidirectional or bidirectional control. Additional robot steering configurations are possible.

3.3.1.1 Mecanum Wheels

Mecanum wheels provide additional flexibility in robot steering. As shown in Fig. 3.6a, Mecanum wheels consist of rollers mounted at a 45° angle from the wheel plane. With four bidirectional, independently controlled, Mecanum wheels; a robot is equipped to make abrupt turns not possible with other configurations as shown in Fig. 3.6b. Due to their specific roller orientation, Mecanum wheels are available in left and right hand configurations (Diegel et al. 2002).

3.3.1.2 Tracked Robots

Figure 3.7 demonstrates how tracked robots are steered. Each track is powered by a bi–directional motor. To render a turn, one track is moved in the forward direction while the other is reversed (McComb 2014).

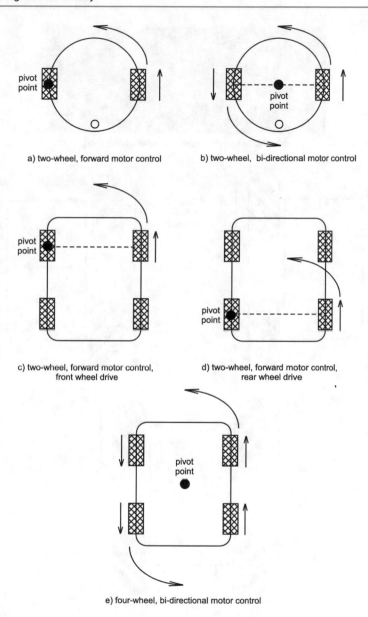

a) two-wheel, forward motor control   b) two-wheel, bi-directional motor control

c) two-wheel, forward motor control,   d) two-wheel, forward motor control,
   front wheel drive   rear wheel drive

e) four-wheel, bi-directional motor control

Fig. 3.5 Robot control configurations

a) Mecanum wheel [images courtesy of adafruit.com].

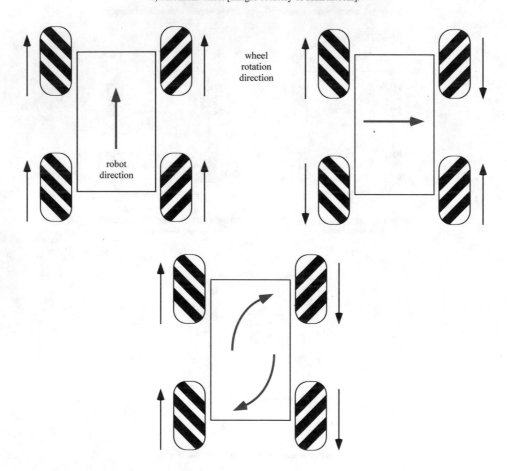

b) steering with Mecanum wheels [Diegel].

Fig. 3.6 Mecanum wheels (Diegel et al. 2002) [6]

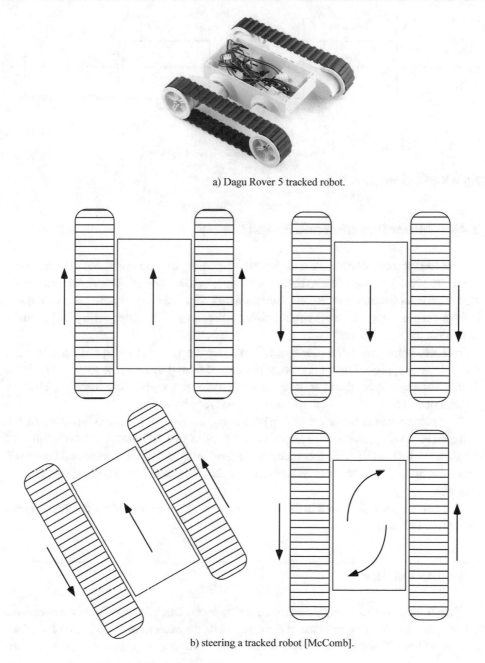

a) Dagu Rover 5 tracked robot.

b) steering a tracked robot [McComb].

Fig. 3.7 Tracked robots (McComb 2014) [14]

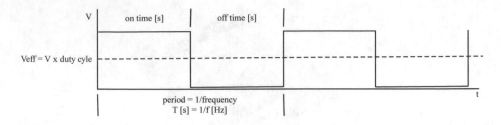

duty cycle [%] = (on time [s] / period [s]) x 100%

Fig. 3.8 PWM concepts

3.4   Motor Direction and Speed Control

To provide for robot steering we must have the capability to control a motor's direction and speed. To provide for bidirectional DC motor control, an H–bridge is typically employed. An H–bridge is an electronic switch configuration that allows the motor power source polarity to be switched using low level control signals. We discuss H–bridge theory and example circuits in the next chapter.

It is also helpful to vary motor speed. A DC motor operates at full speed when its rated DC voltage is applied. When the voltage is removed from the motor, the motor stops. If the supply voltage is applied and removed to a DC motor in a repetitive manner, the effective voltage delivered to the motor and hence its speed may be varied.

The pulse width modulation (PWM) technique allows the precise adjustment of full voltage (on time) versus no voltage (off time) applied to a DC motor. If the on time and off time are both set to 50%, the motor is provided 50% of the full voltage and hence will operate at 50% of its rated speed as shown in Fig. 3.8. We discuss PWM concepts in the next chapter.

The two concepts discussed in this section may be combined to control the direction and speed of a DC motor.

3.5   Odometry

In Chap. 1 we provided a simple control algorithm for the Dagu 5 robot. Turns were rendered by stopping the motor on one side of the robot while the other motor was left on for a fixed amount of time. While this technique is elegant in its simplicity, it has several drawbacks:

–   What happens if the powered wheel or track slips on the surface?
–   What happens if the robot is blocked by a stationary object?
–   What happens if the robot malfunctions and the turn is not rendered properly?

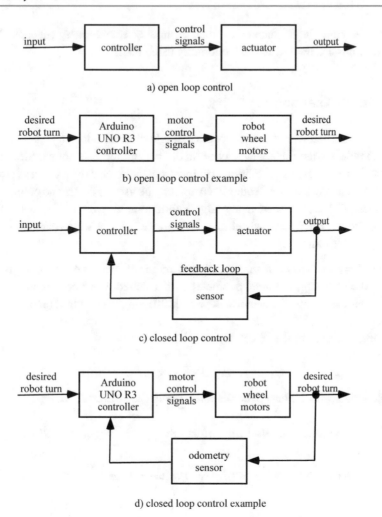

Fig. 3.9 Open versus closed loop control

– What happens if we desire precise, repeatable turns?

The algorithm provided in Chap. 1 may be categorized as open loop control. That is, we have provided control signals to the robot to render the turn but do not receive any feedback the turn was actually accomplished. To provide closed loop control, we need feedback via sensors to indicate the turn has been rendered as desired as shown in Fig. 3.9.

Odometry provides the feedback information necessary to determine if the robot turn has been rendered as desired. Also, wheel odometry provides distance traveled per wheel. Tracking this information allows repeatable robot turns, measurement of distance traveled, and robot velocity.

We next explore single channel and dual channel wheel odometry concepts. We follow up with a discussion on how to track odometry using interrupts.

3.5.1   Single Channel Odometry

Figure 3.10a illustrates the basic concept of single channel odometry. A slotted rotating wheel is attached to the robot wheel axle. An optical emitter–detector pair detects the slots in the wheel as it rotates. The conditioned output from the detector provides a microcontroller compatible signal. A microcontroller interrupt may be used to detect and count the signal's rising edges. The accumulated rising edge information may be used to determine wheel distance traveled. Each robot wheel may be equipped with a single channel odometer to render repeatable turns.

Example: Suppose a robot is equipped with 6.5 cm diameter wheels. Each robot wheel is equipped with a single channel odometer with a slotted wheel containing 20 slots. This means the odometer optical detector will provide 20 rising signal edges for each robot wheel rotation.

The circumference of the wheel is:

$$circumference \ = \ 2 \ \times \ \pi \ \times \ radius$$

Each encoder wheel interrupt corresponds to distance traveled:

$$distance \ traveled \ per \ interrupt \ = \ circumference/slots$$

$$distance \ traveled \ per \ interrupt \ = \ 2 \ \times \ \pi \ \times \ radius/slots$$

$$distance \ traveled \ per \ interrupt \ = \ \pi \ \times \ diameter/slots$$

$$distance \ traveled \ per \ interrupt \ = \ \pi \ \times \ 6.5/20$$

$$distance \ traveled \ per \ interrupt \ = \ 1.02 \ cm/interrupt$$

Example. In this example we equip a Dagu DG007 Magician (Jameco #2210001) two–wheeled robot with an optical encoder as shown in Fig. 3.11. We use a Hitachi HC–020K photoelectric encoder kit to equip the robot wheels with an optical encoder.

Electronically, the encoder is quite easy to use. It has three signal pins (5 VDC, Ground, and Out). The output signal from the encoder is shown in Fig. 3.11c. The rising edge signals may be counted using interrupts.

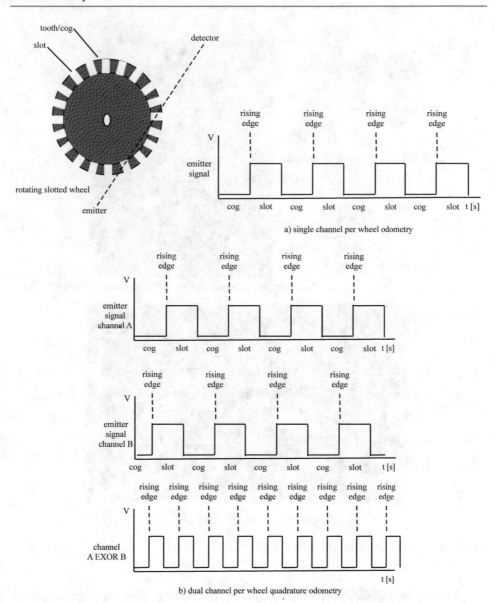

a) single channel per wheel odometry

b) dual channel per wheel quadrature odometry

Fig. 3.10 Wheel odometry

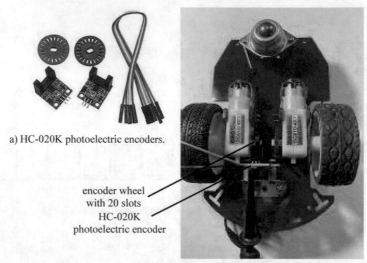

a) HC-020K photoelectric encoders.

encoder wheel
with 20 slots
HC-020K
photoelectric encoder

b) Dagu robot equipped with HC-020K photoelectric encoder.

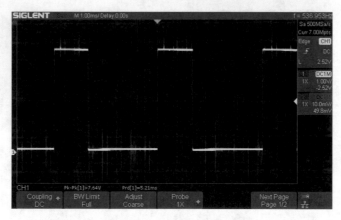

c) HC-020K photoelectric encoder output.

Fig. 3.11 Optical wheel odometry

Mounting the HC–020K photoelectric encoder kit may be challenging depending on the robot design. As discussed in Chap. 2, 3D printing techniques may be used to design and render a custom mounting bracket.

In the sketch below, the output from the HC–020K photoelectric encoder is provided to the INT1 pin on the Arduino UNO R3. The sketch counts the rising signal edges from the HC–020K photoelectric encoder output. The count is used to determine distance traveled by the robot wheel the encoder is monitoring.

```
//*****************************************************************
//travel_distance:
//Program measures the accumulated distance travelled by robot
//wheel.  Positive edge signals from an optical encoder are
//provided to INT1 (pin 3) of the UNO R3.
//*****************************************************************

unsigned int wheel_int = 0;
float distance_trav;
float wheel_diam_cm = 6.5;
int    slots_per_encoder_wheel = 20;

void setup()
{
Serial.begin(9600);
pinMode(3, INPUT);
attachInterrupt(1, int1_ISR, RISING);
}

void loop()
{

//wait for interrupts

}

//*****************************************************************
//int1_ISR: interrupt service routine for INT1
//distance = ((PI x diameter)/slots per encoder wheel) x int count
//*****************************************************************

void int1_ISR(void)
{
wheel_int++;                                   //increment wheel count
                                               //closures to cm
distance_trav = (float)(wheel_int) * ((3.14 * wheel_diam_cm)/
                (float)(slots_per_encoder_wheel));
Serial.print(distance_trav);
Serial.println("    cm");
Serial.println();
}

//*****************************************************************
```

An HC–020K photoelectric encoder is required for each robot powered wheel. Similarly, an interrupt channel is required for each encoder. Turns for a given robot may be calibrated to render repeatable turns using interrupt counts.

3.5.2   Dual–Channel, Quadrature Odometry

A quadrature encoder is equipped with two odometry channels producing signals 90° out of phase with one another as shown in Fig. 3.12. The phased signals allow the measure of distance and direction traveled. Figure 3.12 shows the quadrature encoder onboard the Dagu Rover series robots (ROV–2 and ROV–3) and the resulting quadrature encoder output signals.

As the monitored wheel turns in one direction, the resulting output sequence from the encoder is 00–10–11–01–00. When the wheel rotates in the opposite direction the reverse sequence is obtained.

a) quadrature encoder [www.dagu.com]

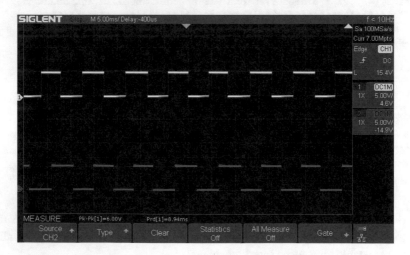

b) quadrature encoder output signal

Fig. 3.12 Quadrature encoder odometry. **a** Quadrature encoder onboard the Dagu Rover series robots (ROV–2 and ROV–3), **b** quadrature encoder signals (www.dagurobot.com) [12]

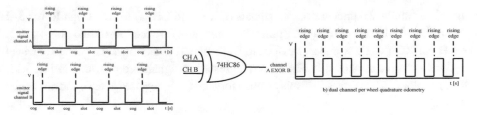

a) CH A EXOR CH B provides a signal with rising edges indicating a change in wheel state.

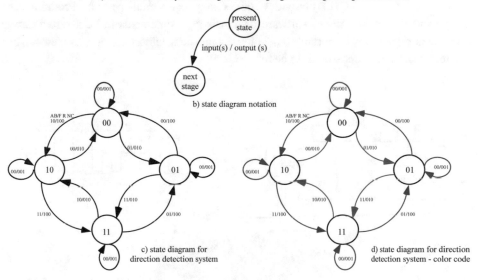

Fig. 3.13 Quadrature encoder state diagram

As shown in Fig. 3.13a, the two encoder channel signals (Channel A and Channel B) may be EXORed together. The resulting signal's edge (rising or falling) may be used to trigger an interrupt indicating a change in wheel status has occurred.

To help develop a sketch to determine wheel direction and distance travelled, a state diagram may be used. State diagram notation is provided in Fig. 3.13b. A state is shown with a circle. A directed arc shows the transition to another state when the specified input is provided. The resulting output signals are generated. The CH A EXOR CH B signal is used to trigger an interrupt and signal the change in state. The resulting state diagram is provided in Fig. 3.13c. A color coded version is provided in Fig. 3.13d. The red arcs illustrate the forward direction of the wheel while the green arcs the reverse direction. The blue arcs provide for a state transition when the wheel is stationary.

Example. In Chap. 1 we equipped a Dagu Rover 5–1 with a control system to avoid obstacles. The "5–1" model is not equipped with odometry encoders. However, the Dagu Rover 5–2 and Rover 5–3 are equipped with two motors and encoders and four motors and encoders respectively.

In this example, the encoder output signals (CH A and CH B) from a Dagu Rover 5–2 robot equipped with wheels (Dagu 28 mm with 4 mm hubs) are fed to Arduino pins 4 and 5. The CH A EXOR CH B signal is provided to INT1 (pin 3) to mark when a change in wheel status has occurred as shown in Fig. 3.14. The resulting signals are provided in Fig. 3.15. Different color LEDs are used to display status: forward (red), reverse (green), and no change (yellow).

The sketch below implements the state diagram provided in Fig. 3.12c and d. The interrupt signal (CH A EXORed CH B) provides 166.5 rising edge signals per wheel rotation. The "troubleshoot" variable allows software diagnosis type statements to be asserted during code development. It is important to note the diagnostic troubleshooting statements, when asserted, causes interrupt signals to be missed.

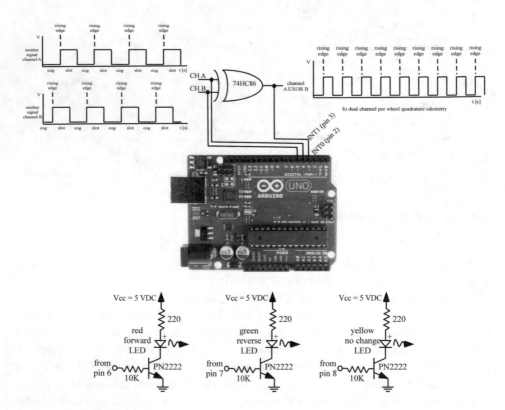

CH A EXOR CH B provides a signal with rising edges indicating a change in wheel state.

Fig. 3.14 Quadrature encoder odometry circuity. UNO R3 illustration used with permission of the Arduino Team (CC BY–NC–SA) (www.arduino.cc) [3]

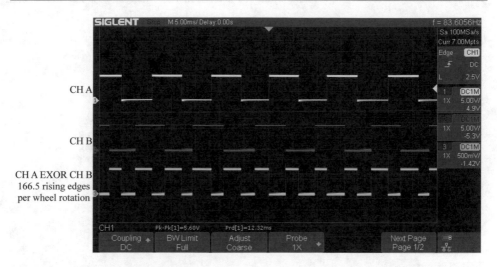

CH A

CH B

CH A EXOR CH B
166.5 rising edges
per wheel rotation

Fig. 3.15 Quadrature encoder signals

```
//******************************************************************
//quad_encoder:
//Program measures the accumulated distance traveled by robot
//wheel.  Positive edge signals (CH A and CH B) from a quadrature
//optical encoder are EXORed to form interrupt signal. The signal
//is provided to INT1 (pin 3) of the UNO R3.
//******************************************************************

unsigned int wheel_int = 0;
float distance_trav;
float wheel_diam_cm = 5.25;                //Dago ROV 5.25 cm wheel diam
int   slots_per_encoder_wheel = 166.5;     //measured: 166.5 per rotation
int   CHA_quad_signal = 4;                 //CHA quadrature signal
int   CHB_quad_signal = 5;                 //CHB quadrature signal
int   red_fwd_LED = 6;                     //fwd LED connected to pin 6
int   grn_rev_LED = 7;                     //rev LED connected to pin 7
int   ylw_nc_LED  = 8;                     //nc  LED connected to pin 8
int   CHA_val, CHB_val;                    //quadrature channel readings
int   previous_state = 0;                  //holds value of previous state
int   troubleshoot = 0;

void setup()
{
Serial.begin(9600);
pinMode(3, INPUT);
pinMode(CHA_quad_signal, INPUT);
pinMode(CHB_quad_signal, INPUT);
pinMode(red_fwd_LED, OUTPUT);
pinMode(grn_rev_LED, OUTPUT);
pinMode(ylw_nc_LED, OUTPUT);
attachInterrupt(1, int1_ISR, RISING);
}
```

```
void loop()
{

//wait for interrupts

}

//*********************************************************************
//int1_ISR: interrupt service routine for INT1
//distance = ((PI x diameter)/slots per encoder wheel) x int count
//*********************************************************************

void int1_ISR(void)
{
CHA_val = digitalRead(CHA_quad_signal);        //read encoder chs
CHB_val = digitalRead(CHB_quad_signal);

if((CHA_val == LOW) && (CHB_val == LOW))      //current state 00
  {
  if(troubleshoot){Serial.println("State:00");  Serial.println();}
  if(previous_state == 1)                     //01
    {
    digitalWrite(red_fwd_LED, HIGH);
    digitalWrite(grn_rev_LED, LOW);
    digitalWrite(ylw_nc_LED,  LOW);
    if(troubleshoot) Serial.println("FWD");
    if(troubleshoot) Serial.println();
    }
  else if (previous_state == 0)               //00 - no change
    {
    digitalWrite(red_fwd_LED, LOW);
    digitalWrite(grn_rev_LED, LOW);
    digitalWrite(ylw_nc_LED,  HIGH);
    if(troubleshoot) Serial.println("NC");
    if(troubleshoot) Serial.println();
    }
  else
    {
    digitalWrite(red_fwd_LED, LOW);
    digitalWrite(grn_rev_LED, HIGH);
    digitalWrite(ylw_nc_LED,  LOW);
    if(troubleshoot) Serial.println("REV");
    if(troubleshoot) Serial.println();
    }
  previous_state = 0;                         //update
  }
else if ((CHA_val == HIGH) && (CHB_val == LOW)) //current state 10
  {
  if(troubleshoot) {Serial.println("State:10");  Serial.println();}
  if(previous_state == 0)                     //00
    {
    digitalWrite(red_fwd_LED, HIGH);
    digitalWrite(grn_rev_LED, LOW);
    digitalWrite(ylw_nc_LED,  LOW);
    if(troubleshoot) Serial.println("FWD");
```

```
     if(troubleshoot) Serial.println();
     }
   else if (previous_state == 10)               //10 - no change
     {
     digitalWrite(red_fwd_LED, LOW);
     digitalWrite(grn_rev_LED, LOW);
     digitalWrite(ylw_nc_LED,  HIGH);
     if(troubleshoot) Serial.println("NC");
     if(troubleshoot) Serial.println();
     }
   else
     {
     digitalWrite(red_fwd_LED, LOW);
     digitalWrite(grn_rev_LED, HIGH);
     digitalWrite(ylw_nc_LED,  LOW);
     if(troubleshoot) Serial.println("REV");
     if(troubleshoot) Serial.println();
     }
   previous_state = 10;                         //update
   }
else if ((CHA_val == HIGH) && (CHB_val == HIGH)) //current state 11
   {
   if(troubleshoot) {Serial.println("State:11");  Serial.println();}
   if(previous_state == 10)                     //10
     {
     digitalWrite(red_fwd_LED, HIGH);
     digitalWrite(grn_rev_LED, LOW);
     digitalWrite(ylw_nc_LED,  LOW);
     if(troubleshoot) Serial.println("FWD");
     if(troubleshoot) Serial.println();
     }
   else if (previous_state == 11)               //11 - no change
     {
     digitalWrite(red_fwd_LED, LOW);
     digitalWrite(grn_rev_LED, LOW);
     digitalWrite(ylw_nc_LED,  HIGH);
     if(troubleshoot) Serial.println("NC");
     if(troubleshoot) Serial.println();
     }
   else
     {
     digitalWrite(red_fwd_LED, LOW);
     digitalWrite(grn_rev_LED, HIGH);
     digitalWrite(ylw_nc_LED,  LOW);
     if(troubleshoot) Serial.println("REV");
     if(troubleshoot) Serial.println();
     }
   previous_state = 11;                         //update
   }
else if ((CHA_val == LOW) && (CHB_val == HIGH)) //current state 01
   {
   if(troubleshoot) {Serial.println("State:10");  Serial.println();}
   if(previous_state == 11)                     //11
     {
     digitalWrite(red_fwd_LED, HIGH);
```

```
      digitalWrite(grn_rev_LED, LOW);
      digitalWrite(ylw_nc_LED,  LOW);
      if(troubleshoot) Serial.println("FWD");
      if(troubleshoot) Serial.println();
      }
  else if (previous_state == 1)              //01 - no change
      {
      digitalWrite(red_fwd_LED, LOW);
      digitalWrite(grn_rev_LED, LOW);
      digitalWrite(ylw_nc_LED,  HIGH);
      if(troubleshoot) Serial.println("NC");
      if(troubleshoot) Serial.println();
      }
  else
      {
      digitalWrite(red_fwd_LED, LOW);
      digitalWrite(grn_rev_LED, HIGH);
      digitalWrite(ylw_nc_LED,  LOW);
      if(troubleshoot) Serial.println("REV");
      if(troubleshoot) Serial.println();
      }
  previous_state = 1;                        //update
  }

wheel_int++;                                 //increment wheel count
                                             //closures to cm
distance_trav = (float)(wheel_int) * ((3.14 * wheel_diam_cm)/
                (float)(slots_per_encoder_wheel));

if (troubleshoot)
  {
  Serial.print(wheel_int);
  Serial.println("   ints");
  Serial.println();

  Serial.print(distance_trav);
  Serial.println("   cm");
  Serial.println();
  }
}

//****************************************************************
```

It is important to note that an interrupt input is required for each quadrature encoder equipped wheel.

3.6   Vision and Obstacle Avoidance

For a robot to navigate about an environment it must have the capability to sense and avoid obstacles. There are several technologies available to provide this capability including:

- infrared sensors,
- ultrasound sensors,
- light detection and ranging or LIDAR, and
- image processing techniques.

Image processing techniques will not be discussed further here. Image processing obstacle detection and avoidance requires more computational horsepower than what is available from the Arduino UNO R3 and the ATmega2560. We discuss each of the remaining techniques in turn.

3.6.1   Infrared Sensor

An infrared sensor consists of an emitter (an infrared LED) and a detector to measure distance. In Chap. 1 we equipped the Dagu Rover 5 robot platform with three Sharp GP2Y0A41SK0F (we abbreviated as GP2Y) infrared (IR) sensors. This sensor measures distances from 4 to 30 cm and provides a corresponding analog DC voltage (Sharp) [27].

Beyond a peak voltage at a sensed distance of 5 cm, the IR sensor provides a voltage output that is inversely proportional to the sensor distance from a reflecting surface as shown in Fig. 3.16. The profile may be experimentally measured by measuring the IR sensor output voltage a various reflection ranges.

Fig. 3.16 IR sensor profile

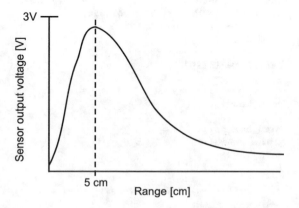

Sharp notes the detector does not provide an accurate reading in the presence of "noise" light sources (e.g. the sun).

3.6.2   Ultrasound Sensor

The ultrasonic sensor pictured in Fig. 3.17 is based on the concept of ultrasound or sound waves that are at a frequency above the human range of hearing (20 Hz to 20 kHz). The ultrasonic sensor pictured in Fig. 3.17c emits a sound wave at 42 kHz. The sound wave reflects from a solid surface and returns back to the sensor. The amount of time for the sound wave to transit to the surface and back to the sensor may be used to determine the range from the sensor to the wall.

Pictured in Fig. 3.17c and d is an ultrasonic sensor manufactured by Maxbotix, the LV-–Max Sonar—EZ series, MB1010–000 [15]. The sensor provides an output that is linearly related to range in three different formats: (a) a serial RS–232 compatible output at 9600 bits per second, (b) a pulse width modulated (PWM) output at a 147 us/in. duty cycle, and (c) an analog output at a resolution of approximately 10 mV/in. when Vcc is 5.0 VDC. The sensor is powered from a 2.5 to 5.5 VDC source [www.sparkfun.com].

Maxbotix provides sample code to test the sensor. In this example we use the sensor's analog output to determine range.

```
//************************************************************
//ultrasonic - demonstration of Maxbotix LV-MaxSonar-EZ
//            MB1010
//
//Source: \url{www.maxbotix.com}
//************************************************************

const int anPin = 0;
int       anVolt;                       //from ultrasound sensor
float     inches, sensor_voltage;

void setup()
{
Serial.begin(9600);
}

void loop()
{
read_sensor();
print_range();
delay(100);
}

//************************************************************
void read_sensor()
{
anVolt = analogRead(anPin);            //returns 0 to 1023
```

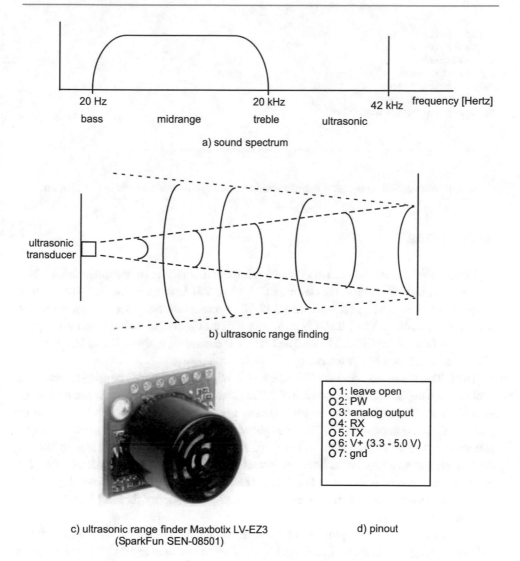

20 Hz   20 kHz   42 kHz   frequency [Hertz]

bass   midrange   treble   ultrasonic

a) sound spectrum

ultrasonic
transducer

b) ultrasonic range finding

O 1: leave open
O 2: PW
O 3: analog output
O 4: RX
O 5: TX
O 6: V+ (3.3 - 5.0 V)
O 7: gnd

c) ultrasonic range finder Maxbotix LV-EZ3   d) pinout
(SparkFun SEN-08501)

Fig. 3.17 Ultrasonic sensor. Sensor image used courtesy of SparkFun Electronics (CC BY–NC–SA) (www.sparkfun.com)

```
sensor_voltage = (float(anVolt)/1023.0) * 5.0;    //to voltage
inches = sensor_voltage/0.009766;                 //to inches
}

//*********************************************************

void print_range()
{
Serial.print("Range:");
```

```
Serial.print(" ");
Serial.print(anVolt);
Serial.print(" ");
Serial.print(sensor_voltage);
Serial.print(" ");
Serial.print(inches);
Serial.println(" inches");
}

//***********************************************************
```

Additional applications using the ultrasonic sensor are provided later in the chapter.

3.6.3   LIDAR

Light detection and ranging or LIDAR allows a robot to sense its environment using a low power laser beam. We explore the Garmin LIDAR Lite V3 (Adafruit #4058) [9]. This LIDAR range finder, shown in Fig. 3.18a uses a 1.3 W, 905 nm near–infrared laser to range objects at a distance up to 40 m. The LIDAR is stationary so it provides range information for objects directly in front of the LIDAR head. The LIDAR is connected to the Arduino UNO R3 via either via an I2C or a PWM interface.

The LIDAR is tested using the I2C interface configuration. LIDAR ranging information is obtained using sketch "GetDistanceI2C." This sketch is available in the Arduino LIDAR-–Lite Library. The LIDAR may be connected to a Dagu HiTech Electronic RS002B sensor pan and tilt kit (Jameco #2157870) [5] as shown in Fig. 3.18b. This allows the LIDAR to pan a room. The kit is equipped with two servo motors to provide for the pan and tilt feature. An operational amplifier based interface circuit is required between the Arduino UNO R3 and the servo motors as shown in Fig. 3.18c. The assembled test rig is shown in Fig. 3.18d. It is highly recommended to use an external 5 VDC power supply for the LIDAR Lite V3 and the interface circuit.

Two sketches may be employed to test the servos and then integrate the servos with the LIDAR ranging unit. In the first sketch the X and Y servos are used to provide control signals for the pan and tilt mechanism.

```
//***************************************************************
//X-Y ramp: provides pan and tilt control signals.
//
//Note: requires LM324 op amp based interface circuit between
//      UNO R3 and pan and tilt servos
//***************************************************************
#include <Servo.h>        //Use Servo library, included with IDE
Servo myServo_x;          //Create Servo object to control the servo
Servo myServo_y;

void setup()
```

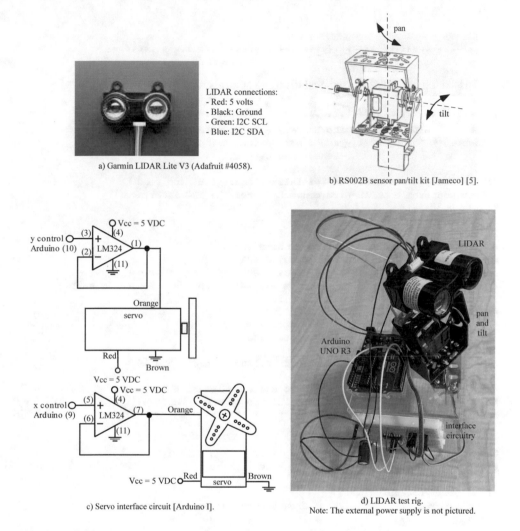

a) Garmin LIDAR Lite V3 (Adafruit #4058).

LIDAR connections:
- Red: 5 volts
- Black: Ground
- Green: I2C SCL
- Blue: I2C SDA

b) RS002B sensor pan/tilt kit [Jameco] [5].

c) Servo interface circuit [Arduino I].

d) LIDAR test rig.
Note: The external power supply is not pictured.

Fig. 3.18 LIDAR Lite V3. Image used courtesy of (www.adafruit.com)

```
{
mmyServo_x.attach(9);    //Servo is connected to digital pin 9
myServo_y.attach(10);    //Servo is connected to digital pin 10
}
void loop()
{
int i = 0;
for(i=50; i<=150; i++)
  {
  //myServo_x.write(i);//Rotate servo counter clockwise
  myServo_y.write(i);    //Rotate servo counter clockwise
  delay(500);            //Wait 200 milliseconds
```

```
  }
}
//****************************************************************
```

This sketch integrates the pan and tilt servo controls with the LIDAR unit.

```
//****************************************************************
//Source:  LIDARLite Arduino Library - 3/GetDistanceI2C
//Adapted:  May 8, 2021 - added pan feature using RS002B
//                        sensor pan/tilt kit
//
//This example shows how to initialize, configure, and read
//distance from a LIDAR-Lite connected over the I2C interface.
//
//Connections:
//LIDAR-Lite 5 Vdc (red) to Arduino 5v
//LIDAR-Lite I2C SCL (green) to Arduino SCL
//LIDAR-Lite I2C SDA (blue) to Arduino SDA
//LIDAR-Lite Ground (black) to Arduino GND
//Capacitor recommended to mitigate inrush current when device is enabled
//   680uF capacitor (+) to Arduino 5v
//   680uF capacitor (-) to Arduino GND
//   X servo - pin  9 with LM324 interface
//   Y servo - pin 10 with LM324 interface
//
//Note: See the Operation Manual for wiring diagrams and more
//      information.
//Note: It is highly recommended to use an external 5 VDC power
//      supply for the LIDAR unit and the op amp interface.
//****************************************************************

#include <Wire.h>
#include <LIDARLite.h>
#include <Servo.h>             //Use Servo library

LIDARLite myLidarLite;
Servo myServo_x;               //Create servo objects
Servo myServo_y;

void setup()
{
Serial.begin(115200);          //Init serial comm - distance readings
myLidarLite.begin(0, true);    //Set config to default,  I2C to 400 kHz
myLidarLite.configure(0);      //Adjust for alternate configurations
myServo_x.attach(9);           //Servo is connected to digital pin 9
myServo_y.attach(10);          //Servo is connected to digital pin 10
}

void loop()
{
int i = 0;
for(i=50; i<=150; i++)         //define servo span for pan motion
  {
  //myServo_x.write(i);
```

```
  myServo_y.write(i);        //Rotate servo counter clockwise
                             //LIDAR measurement
  Serial.println(myLidarLite.distance());
  delay(100);                //Wait 100 milliseconds
  }
}


//********************************************************************
```

3.7   Aside: TFT Display

We take a pause from our robot feature discussion to present an aside on advanced display technology. In previous volumes of this Arduino series we have used Liquid Crystal Displays (LCDs) to report microcontroller status, sensor output, etc. We explore TFT displays which provide great flexibility in displaying graphics, text, sensor readings, etc.

A TFT display consists of individually addressable picture elements (pixels). A pixel's location is specified with an X, Y coordinate as shown in Fig. 3.19. The color for an individual pixel is set by specifying the amount of red, green, and blue color components. Graphics, texts, and images are a collection of related pixels [Ada TFT] [1].

We investigate the Adafruit ILI9341 full color TFT display. This display has a 2.8 in. diagonal dimension and features 240 by 320 pixel resolution with 18–bit color. The display is available in a fully configured Arduino UNO R3 compatible shield. The communication

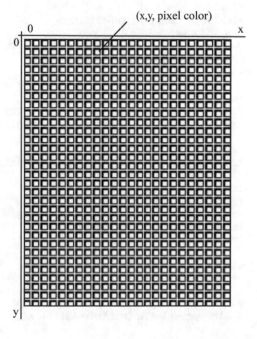

Fig. 3.19 TFT display coordinate system

between the UNO R3 and the display is via a Serial Peripheral Interface (SPI) employing Arduino UNO R3 pins 9 through 13. There are several application libraries available for the TFT display (Ada TFT 2020). The libraries are downloaded using the Arduino Library Manager.

To get acquainted with the display, we use an extended example of an "old school" sonar display. If you watched old movies of ship or submarine sonar, you will recognize our display. It consists of a green circle with the sonar sensor in the center. Ranging information is provided as a displacement from display center. In this example we use the ultrasound sensor introduced earlier in the chapter, equip it with a pan and tilt mechanism, a sonar TFT "green screen" display, and also an audio signal. We implement this project in stages beginning with the sonar graphics display as shown in Fig. 3.20b.

Adafruit provides a GFX Graphics Library to support the TFT display [4]. It provides a wide variety of functions to draw common graphic features such as lines, circles, text, etc. In the following example, we adapted the Adafruit provided "graphicstest" sketch to provide the sonar test pattern.

```
//**********************************************************************
//Sonar
//Adapted from Adafruit GFX example
//Written by Limor Fried/Ladyada for Adafruit Industries.
//MIT license, all text above must be included in any redistribution
//Adapted by:  S. Barrett, May 10, 2021
//**********************************************************************

#include "SPI.h"
#include "Adafruit_GFX.h"
#include "Adafruit_ILI9341.h"
#include <math.h>

#define TFT_DC 9                              //Adafruit shield defaults
#define TFT_CS 10
                                              //global variables
int     x1, y1, x2, y2, screen_ctr_x, screen_ctr_y;
int     width, height, max_range = 120;
int     angle_int, angle_min = 1, angle_max = 179, angle_inc = 1;

                                              //Hardware UNO SPI:#13, #12, #11
Adafruit_ILI9341 tft = Adafruit_ILI9341(TFT_CS, TFT_DC);

void setup()
{
Serial.begin(9600);
Serial.println("ILI9341 Test!");
tft.begin();

width = tft.width(),
height = tft.height();

Serial.print(F("Screen fill              "));
Serial.println(testFillScreen());
```

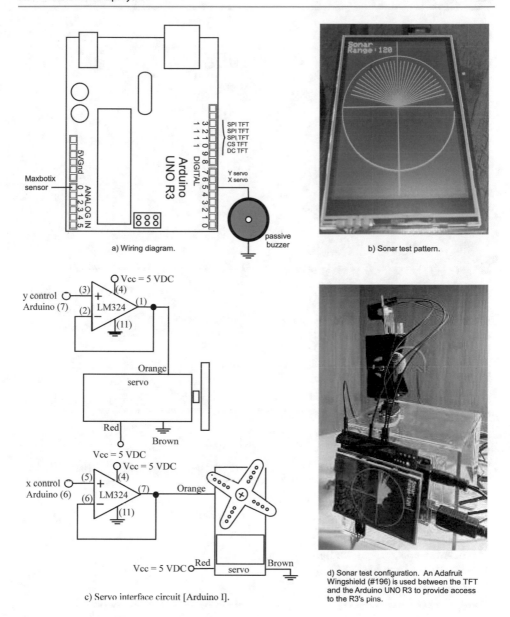

a) Wiring diagram.

b) Sonar test pattern.

c) Servo interface circuit [Arduino I].

d) Sonar test configuration. An Adafruit Wingshield (#196) is used between the TFT and the Arduino UNO R3 to provide access to the R3's pins.

Fig. 3.20 Sonar simulator using Maxbotix sensor

```
delay(500);

Serial.print(F("Lines                    "));
Serial.println(testLines(ILI9341_GREEN));
delay(500);

Serial.print(F("Text                     "));
Serial.println(testText());
delay(3000);
}


void loop(void)
{
//insert ranging code

}

//**********************************************************************

unsigned long testFillScreen()
{
tft.fillScreen(ILI9341_BLACK);
unsigned long start = micros();
tft.fillScreen(ILI9341_BLACK);
yield();
return micros() - start;
}

//**********************************************************************

unsigned long testText()
{
unsigned long start = micros();

tft.setCursor(0, 0);
tft.setTextColor(ILI9341_GREEN);
tft.setTextSize(2);
tft.println("Sonar");
tft.print("Range:");
tft.println(max_range, DEC);
return micros() - start;
}

//**********************************************************************

unsigned long testLines(uint16_t color)
{
unsigned long start, t;

start = micros();

//calculate screen center
screen_ctr_x = width/2;    screen_ctr_y = height/2;
```

```
//draw crosshairs
tft.drawLine(screen_ctr_x, 0, screen_ctr_x, height, ILI9341_GREEN);
tft.drawLine(0, screen_ctr_y, width, screen_ctr_y, ILI9341_GREEN);

//draw circle
tft.drawCircle(screen_ctr_x, screen_ctr_y, screen_ctr_x, ILI9341_GREEN);

//draw range lines
for(angle_int=angle_min;angle_int<=angle_max;angle_int=angle_int+angle_inc)
  {
  x2=(int)((float)(max_range)*0.98*cos((float)(angle_int)*3.14/180.0));
  y2=(int)((float)(max_range)*0.98*sin((float)(angle_int)*3.14/180.0));
  tft.drawLine(screen_ctr_x,screen_ctr_y,screen_ctr_x+x2,
               screen_ctr_y-y2, ILI9341_GREEN);
  }

t = micros() - start; // fillScreen doesn't count against timing
yield();
return micros() - start;
}


//********************************************************************
```

In the next step, we equip the Maxbotix LV–MaxSonar–EZ sensor with a pan and tilt mechanism discussed earlier in the chapter.

```
//********************************************************************
//GetDistance_MaxBotix
//Adapted from \url{www.maxbotox.com}
//ultrasonic - demonstration of Maxbotix LV-MaxSonar-EZ
//               MB1010
//Adapted:  May 16, 2021 - added pan feature using RS002B
//                         sensor pan/tilt kit
//  X servo - pin  6 with LM324 interface
//  Y servo - pin  7 with LM324 interface
//
//********************************************************************

#include <Servo.h>       //Use Servo library, included with IDE

Servo myServo_x;         //Create Servo object to control the servo
Servo myServo_y;

const int anPin = 0;
int       anVolt;        //from ultrasound sensor
float     inches, sensor_voltage;


void setup()
{
Serial.begin(9600);      //Init serial comm - distance readings
myServo_x.attach(6);     //Servo is connected to digital pin 9
myServo_y.attach(7);     //Servo is connected to digital pin 10
}
```

```
void loop()
{
int i = 0;
for(i=50; i<=150; i++)   //define servo span
  {
  myServo_x.write(0);    //Rotate servo counter clockwise
  myServo_y.write(i);    //Rotate servo counter clockwise
  read_sensor();
  print_range();
  delay(100);
  }
}


//*********************************************************************

void read_sensor()
{
anVolt = analogRead(anPin);                //returns 0 to 1023
sensor_voltage = (float(anVolt)/1023.0) * 5.0;   //to voltage
inches = sensor_voltage/0.009766;                //to inches
}


//*********************************************************************

void print_range()
{
Serial.print("Range:");
Serial.print(" ");
Serial.print(anVolt);
Serial.print(" ");
Serial.print(sensor_voltage);
Serial.print(" ");
Serial.print(inches);
Serial.println(" inches");
}


//*********************************************************************
```

In the last step, we integrate the sonar display with the pan and tilt control.

```
//*********************************************************************
//GetDistance_MaxBotix_TFT
//Adapted from \url{www.maxbotox.com}
//ultrasonic - demonstration of Maxbotix LV-MaxSonar-EZ
//              MB1010
//Adapted:  May 18, 2021 - added pan feature using RS002B
//                          sensor pan/tilt kit
//                        - added TFT display
//   X servo - pin  6 with LM324 interface
//   Y servo - pin  7 with LM324 interface
//
//   TFT DC  - pin 9
//   TFT CS - pin 10
```

```
//  TFT SPI - pins 11, 12, 13
//
//  passive buzzer - pin 5    tone generation
//
//***********************************************************************

#include <Servo.h>                    //Use Servo library, included with IDE
#include "SPI.h"                      //Required by TFT display
#include "Adafruit_GFX.h"             //Required by TFT display
#include "Adafruit_ILI9341.h"         //Required by TFT display
#include <math.h>

#define TFT_DC 9                      //Adafruit TFT shield defaults
#define TFT_CS 10
                                          //global variables
int       x1, y1, x2, y2, screen_ctr_x, screen_ctr_y;
int       width, height, max_range_pix = 120, max_range_in = 36;
int       angle_int, angle_min = 1, angle_max = 179, angle_inc = 1;
const int anPin = 0;
int       anVolt;                     //from Maxbotix sensor
float     inches, sensor_voltage;
int       tone_gen = 5;

Adafruit_ILI9341 tft = Adafruit_ILI9341(TFT_CS, TFT_DC);
Servo myServo_x;            //Create Servo object to control the servo
Servo myServo_y;

void setup()
{
Serial.begin(9600);          //Init serial comm - distance readings
myServo_x.attach(6);         //Servo is connected to digital pin 9
myServo_y.attach(7);         //Servo is connected to digital pin 10
pinMode(tone_gen, OUTPUT); //passive buzzer

Serial.println("ILI9341 Test!");
tft.begin();

width = tft.width(),
height = tft.height();

Serial.print(F("Screen fill            "));
Serial.println(testFillScreen());
delay(500);

Serial.print(F("Lines                  "));
Serial.println(testLines(ILI9341_GREEN));
delay(500);

Serial.print(F("Text                   "));
Serial.println(testText());
delay(3000);
}

void loop()
{
```

```
long freq_C4 = 262, freq_D4 = 294, freq_E4 = 300, freq_F4 = 349;
long freq_G4 = 392, freq_A4 = 440, freq_B4 = 494, freq_C5 = 784;

for(angle_int=angle_min;angle_int<=angle_max;angle_int=angle_int+angle_inc)
  {
  myServo_x.write(0);              //Rotate servo counter clockwise
  myServo_y.write(angle_int);    //Rotate servo counter clockwise
  read_sensor();
  print_range();

  //draw black line
  x2 = (int)((float)(max_range_pix) * 0.98 * cos((float)(angle_int)
                              * 3.14/180.0));
  y2 = (int)((float)(max_range_pix) * 0.98 * sin((float)(angle_int)
                              * 3.14/180.0));
  tft.drawLine(screen_ctr_x,screen_ctr_y,
                screen_ctr_x + x2, screen_ctr_y - y2, ILI9341_BLACK);

  //draw green line to range
  x2 = (int)(((float)(inches)/(float)(max_range_in))*
              (float)(max_range_pix) * 0.98 *
               cos((float)(angle_int) * 3.14/180.0));
  y2 = (int)(((float)(inches)/(float)(max_range_in))*
              (float)(max_range_pix) * 0.98 *
               sin((float)(angle_int) * 3.14/180.0));
  tft.drawLine(screen_ctr_x,screen_ctr_y,
                screen_ctr_x + x2, screen_ctr_y - y2, ILI9341_GREEN);
  tone(tone_gen, freq_C4 ); delay(500);

  noTone(tone_gen);
  delay(100);
  }
}

//**********************************************************************

void read_sensor()
{
anVolt = analogRead(anPin);                //returns 0 to 1023
sensor_voltage = (float(anVolt)/1023.0) * 5.0;   //to voltage
inches = sensor_voltage/0.009766;                //to inches
}

//**********************************************************************

void print_range()
{
Serial.print("Range:");
Serial.print(" ");
Serial.print(anVolt);
Serial.print(" ");
Serial.print(sensor_voltage);
Serial.print(" ");
Serial.print(inches);
Serial.println(" inches");
```

```
}

//********************************************************************

unsigned long testFillScreen()
{
tft.fillScreen(ILI9341_BLACK);
unsigned long start = micros();
tft.fillScreen(ILI9341_BLACK);
yield();
return micros() - start;
}

//********************************************************************

unsigned long testText()
{
unsigned long start = micros();

tft.setCursor(0, 0);
tft.setTextColor(ILI9341_GREEN);
tft.setTextSize(2);
tft.println("Sonar");
tft.print("Range:");
tft.println(max_range_pix, DEC);
return micros() - start;
}

//********************************************************************

unsigned long testLines(uint16_t color)
{
unsigned long start, t;

start = micros();

//calculate screen center
screen_ctr_x = width/2;    screen_ctr_y = height/2;

//draw crosshairs
tft.drawLine(screen_ctr_x, 0, screen_ctr_x, height, ILI9341_GREEN);
tft.drawLine(0, screen_ctr_y, width, screen_ctr_y, ILI9341_GREEN);

//draw circle
tft.drawCircle(screen_ctr_x, screen_ctr_y, screen_ctr_x, ILI9341_GREEN);

//draw range lines
for(angle_int=angle_min;angle_int<=angle_max;angle_int=angle_int+angle_inc)
  {
  x2 = (int)((float)(max_range_pix) *
        0.98 * cos((float)(angle_int) * 3.14/180.0));
  y2 = (int)((float)(max_range_pix) *
        0.98 * sin((float)(angle_int) * 3.14/180.0));
  tft.drawLine(screen_ctr_x,screen_ctr_y,
              screen_ctr_x + x2, screen_ctr_y - y2, ILI9341_GREEN);
```

```
    }

t = micros() - start; // fillScreen doesn't count against timing
yield();
return micros() - start;
}


//*****************************************************************
```

3.8   Robot Status

It is an important feature for a robot to locally sense its orientation in the X, Y, Z coordinate system An inertial measurement unit (IMU) provides this capability. An IMU consists of an accelerometer, gyroscope, and magnetometer to measure acceleration, rotation, and the Earth's magnetic field strength in the X, Y, Z coordinate system as shown in Fig. 3.21.

Example: In this example we explore the Adafruit Nine Degree of Freedom (9–DOF) [28] Orientation IMU Fusion BNO085 (#4754) as shown in Fig. 3.22. We use an Arduino Mega 2560 for its additional processing power to display data from the IMU. We then interface a TFT display to display yaw, pitch, and roll of the test platform. A thorough review of the Adafruit IMU documentation is highly recommended before continuing (Siepert 2021).

The interface options between the BNO085 IMU to the Mega 2560 are quite flexible (SPI, I2C, or UART). In this example, we use the I2C interface. To display the IMU data

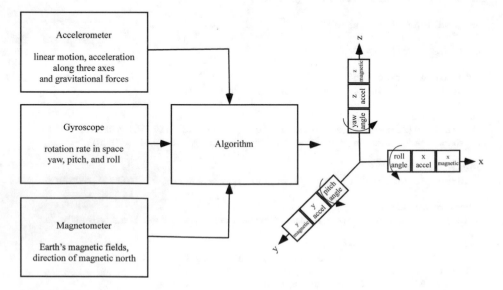

Fig. 3.21 Inertial Measurement Unit

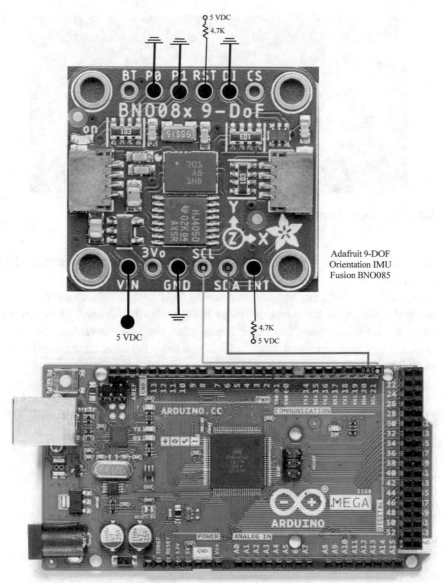

Adafruit 9-DOF IMU with Arduino Mega 2560 I2C configuration [Siepert].

Fig. 3.22 IMU interface to Mega 2560. UNO R3 illustration used with permission of the Arduino Team (CC BY–NC–SA) (www.arduino.cc) [3]. IMU illustration used with permission of Adafruit (www.adafruit.com)

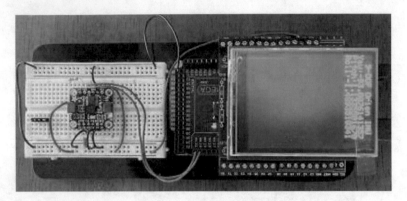

Fig. 3.23 IMU test platform

on the serial monitor, the "quaterion_yaw_pitch_roll" sketch from the Adafruit BNO08x Library is used.

We equip the test platform with the Adafruit ILI 9341 2.8" TFT Touch (resistive) shield. **To use this shield with the Arduino Mega 2560, traces 11–13 need to be cut on the backside of the shield. Also, the ICSP bridge pins must be connected.** The TFT was integrated to the "quaterion_yaw_pitch_roll" sketch using the GFX Library functions. The resulting test platform is shown in Fig. 3.23.

```
//**********************************************************************
//quaterion_yaw_pitch_roll2_tft:  provides Adafruit 9-DOF Orientation IMU
//                                BNO085 data on TFT display
//This demo explores two reports:
//  - SH2_ARVR_STABILIZED_RV
//  - SH2_GYRO_INTEGRATED_RV
//both can be used to give quartenion and euler (yaw, pitch roll) angles.
//Toggle the FAST_MODE define to see other report.
//
// Note sensorValue.status gives calibration accuracy.
//
//Source:  This example was adapted from Adafruit example:
//  - quaterion_yaw_pitch_roll
//
//This example was adapted to provide output on an Adafruit TFT display.
//To use the TFT resistive touch shield with the Arduino Mega 2560, you
//will need to cut TFT traces on 11-13 and solder bridge the ICSP pins
//on the back of the TFT shield.
//  - TFT DC  - pin 9
//  - TFT CS - pin 10
//**********************************************************************

#include <Arduino.h>
#include <Adafruit_BNO08x.h>        //Required for
#include "SPI.h"                    //Required by TFT display
```

```
#include "Adafruit_GFX.h"              //Required by TFT display
#include "Adafruit_ILI9341.h"          //Required by TFT display

// For SPI mode, we need a CS pin
#define BNO08X_CS 10
#define BNO08X_INT 9

//TFT display
#define TFT_DC 9                       //Adafruit TFT shield defaults
#define TFT_CS 10

// #define FAST_MODE

#define BNO08X_RESET -1

struct euler_t
{
  float yaw;
  float pitch;
  float roll;
} ypr;

Adafruit_BNO08x  bno08x(BNO08X_RESET);
sh2_SensorValue_t sensorValue;
Adafruit_ILI9341 tft = Adafruit_ILI9341(TFT_CS, TFT_DC);

#ifdef FAST_MODE
  //Top frequency is reported to be 1000Hz (but freq is somewhat variable)
  sh2_SensorId_t reportType = SH2_GYRO_INTEGRATED_RV;
  long reportIntervalUs = 2000;
#else
  //Top frequency is about 250Hz but this report is more accurate
  sh2_SensorId_t reportType = SH2_ARVR_STABILIZED_RV;
  long reportIntervalUs = 5000;
#endif

void setup(void)
{
Serial.begin(115200);
while (!Serial) delay(10);      //will until serial console opens
Serial.println("Adafruit BNO08x test!");
Serial.println("ILI9341 Test!");
tft.begin();

if(!bno08x.begin_I2C())
  {                               //try to initialize!
  Serial.println("Failed to find BNO08x chip");
  while (1)
    {
    delay(10);
    }
  }
Serial.println("BNO08x Found!");
setReports(reportType, reportIntervalUs);
Serial.println("Reading events");
```

```
delay(100);
}



//*********************************************************************

void quaternionToEuler(float qr, float qi, float qj, float qk,
                       euler_t* ypr, bool degrees = false)
{
float sqr = sq(qr);
float sqi = sq(qi);
float sqj = sq(qj);
float sqk = sq(qk);

ypr->yaw = atan2(2.0 * (qi * qj + qk * qr), (sqi - sqj - sqk + sqr));
ypr->pitch = asin(-2.0 * (qi * qk - qj * qr) / (sqi + sqj + sqk + sqr));
ypr->roll = atan2(2.0 * (qj * qk + qi * qr), (-sqi - sqj + sqk + sqr));

if(degrees)
  {
  ypr->yaw *= RAD_TO_DEG;
  ypr->pitch *= RAD_TO_DEG;
  ypr->roll *= RAD_TO_DEG;
  }
}

//*********************************************************************

void quaternionToEulerRV(sh2_RotationVectorWAcc_t* rotational_vector,
                         euler_t* ypr, bool degrees = false)
{
quaternionToEuler(rotational_vector->real, rotational_vector->i,
                  rotational_vector->j, rotational_vector->k,
                  ypr, degrees);
}

//*********************************************************************

void quaternionToEulerGI(sh2_GyroIntegratedRV_t* rotational_vector,
                         euler_t* ypr, bool degrees = false)
{
quaternionToEuler(rotational_vector->real, rotational_vector->i,
        rotational_vector->j, rotational_vector->k, ypr, degrees);
}

//*********************************************************************

void setReports(sh2_SensorId_t reportType, long report_interval)
{
Serial.println("Setting desired reports");
if(! bno08x.enableReport(reportType, report_interval))
  {
  Serial.println("Could not enable stabilized remote vector");
  }
```

```
}

//**********************************************************************

void loop()
{

if(bno08x.wasReset())
  {
  Serial.print("sensor was reset ");
  setReports(reportType, reportIntervalUs);
  }

delay(1000);                                          //TFT update time

if(bno08x.getSensorEvent(&sensorValue))
  {
  //in this demo only one report type will be received
  //depending on FAST_MODE defined(above)
  switch (sensorValue.sensorId)
    {
    case SH2_ARVR_STABILIZED_RV:
    quaternionToEulerRV(&sensorValue.un.arvrStabilizedRV, &ypr, true);
    case SH2_GYRO_INTEGRATED_RV:
    quaternionToEulerGI(&sensorValue.un.gyroIntegratedRV, &ypr, true);
    break;
    }

  static long last = 0;
  long now = micros();
  Serial.print(now - last);          Serial.print("\t");
  last = now;
  Serial.print(sensorValue.status); Serial.print("\t");
  Serial.print(ypr.yaw);             Serial.print("\t");
  Serial.print(ypr.pitch);           Serial.print("\t");
  Serial.println(ypr.roll);

  //data to TFT
  unsigned long start = micros();
  tft.fillScreen(ILI9341_BLACK);
  yield();

  tft.setCursor(0, 0);
  tft.setTextColor(ILI9341_GREEN);
  tft.setTextSize(2);
  tft.println("  9-DOF Orien IMU");

  tft.setCursor(0, 20);
  tft.setTextColor(ILI9341_GREEN);
  tft.setTextSize(2);
  tft.print("Yaw:");
  tft.println(ypr.yaw, DEC);
  tft.print("Pitch:");
  tft.println(ypr.pitch, DEC);
  tft.print("Roll:");
```

```
    tft.println(ypr.roll, DEC);
    return micros() - start;

    }
}

//**********************************************************************

unsigned long testFillScreen()
{
tft.fillScreen(ILI9341_BLACK);
unsigned long start = micros();
tft.fillScreen(ILI9341_BLACK);
yield();
return micros() - start;
}

//**********************************************************************

unsigned long testText()
{
unsigned long start = micros();

tft.setCursor(0, 0);
tft.setTextColor(ILI9341_GREEN);
tft.setTextSize(2);
tft.println("  9-DOF Orien IMU");
return micros() - start;
}

//**********************************************************************
```

3.8.1   Advanced: Quaternions

One of the series of available outputs from the BNO085 Adafruit 9–DOF Orientation IMU
Fusion is the absolute orientation/rotation vector expressed as a four–point quarternion. The
concept of quaternions is beyond the scope of this book. The interested reader is referred
to "Quaternions and Rotation Sequences–A Primer with Applications to Orbits, Aerospace,
and Virtual Reality" by J. B. Kuipers. This is a very readable introduction to this vital topic
(Kuipers 1999) [13].

3.9   Environmental Sensors

An important application for robots is to go into dangerous areas, not safe for humans,
and sense the environment. In previous volumes of this Arduino series, we have explored
multiple sensors that may be used and interfaced to the Arduino series of processors. In this
section we concentrate on the MQ series of gas and environmental sensors. The MQ series

MQ-2	liquefied petroleum gas (LPG), propane, hydrogen, methane $V_C < 24V$, $V_H = 5V$, $R_H = 31$ ohms, $P_H < 900$ mW, preheat: 48 hours
MQ-3	alcohol, benzene, methane, hexane, LPG, carbon monoxide $V_C = 5V$, $V_H = 5V$, $R_H = 33$ ohms, $P_H < 750$ mW, preheat: 24 hours
MQ-4	high sensitivity to methane and natural gas, lower sensitivity to alcohol and smoke $V_C = 5V$, $V_H = 5V$, $R_H = 33$ ohms, $P_H < 750$ mW, preheat: 24 hours
MQ-6	liquefied petroleum gas (LPG) $V_C < 24V$, $V_H = 5V$, $R_H = 26$ ohms, $P_H < 950$ mW, preheat: 24 hours
MQ-7	carbon monoxide $V_C = 5V$, $R_H = 33$ ohms, $P_H < 350$ mW, preheat: 48 hours V_H : alternates between 5V for 60s and 1.4V for 90s
MQ-8	hydrogen $V_C = 5V$, $V_H = 5V$, $R_H = 29$ ohms, $P_H < 900$ mW, preheat: 48 hours
MQ-9	methane, propane, carbon monoxide $V_C < 10V$, $R_H = 31$ ohms, $P_H < 350$ mW, preheat: 48 hours V_H : alternates between 5V for 60s and 1.5V for 90s
MQ-131	ozone $V_C = 5V$, $V_H = 6V$, $R_H = 31$ ohms, $P_H < 1,100$ mW, preheat: 24 hours
MQ-135	ammonia, nitrogen oxide, alochol, benzene, smoke, carbon dioxide $V_C = 5V$, $V_H = 5V$, $R_H = 33$ ohms, $P_H < 800$ mW, preheat: 24 hours
MQ-138	hexane, benzene, ammonia, alcohol, smoke, carbon monoxide $V_C = 5V$, $V_H = 5V$, $R_H = 31$ ohms, $P_H < 800$ mW, preheat: 24 hours
MQ-214	methane, LPG, butane, propane $V_C = 6V$

Fig. 3.24 MQ sensor series. Data sheets for many of the MQ sensors are provided at www.mysensors. org [16–26]

of sensors consists of a metal oxide semiconductor active element. The electrical resistance of the active element varies in the presence of a specific gas or gases. A sample of available MQ sensors is shown in Fig. 3.24 [www.mysensors.org].

An MQ sensor is typically used in a voltage divider circuit as shown in Fig. 3.25a. In the presence of a specific gas or gases, the resistance of the sensing element (R_S) varies. The value of R_S is in series with a fixed load resistor (R_L) forming a voltage divider network. The output voltage is an indication of the gas concentration. Many of the sensors in the MQ series require a heater voltage as shown in Fig. 3.25b. **Note:** The heater voltage should be provided by an external power supply rather than the Arduino UNO R3. The "R3" does not have sufficient drive current for the heater requirement. Figure 3.25c shows the physical configuration of the MQ series of sensors. SparkFun provides a breakout board to allow an MQ sensor to interface with a standard prototype board [www.mysensors.org, www. SparkFun.com] [29].

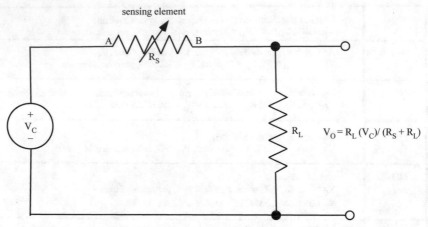

$$V_O = R_L (V_C)/(R_S + R_L)$$

a) electrical resistance (RS) of the sensing element varies in the presence of a specific gas or gases.

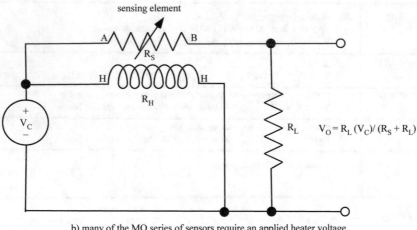

$$V_O = R_L (V_C)/(R_S + R_L)$$

b) many of the MQ series of sensors require an applied heater voltage.

c) MQ sensors and MQ breakout board
[Sensor images courtesy of SparkFun Electronics, Inc. (www.sparkfun.com)].

Fig. 3.25 MQ sensor interface

To develop an interface circuit for an MQ sensor, complete the following steps:

– Choose a specific detectable gas of interest.
– Choose an appropriate sensor from Fig. 3.24.
– Determine key interface parameters as provided in Fig. 3.24.
– Implement the interface circuit as shown in Fig. 3.25b.
– Select an appropriate value of load resistor (R_L). A suggested value of load resistance is provided in the sensor's data sheet.[1] It is recommended to use a potentiometer that includes the value of suggested load resistance as the load resistor. This will allow the adjustment of circuit sensitivity to a specific level of gas concentration.
– Write an Arduino sketch to read the analog output voltage from the sensor interface circuit, set a threshold for detection, and illuminate an LED and sound a buzzer when the gas is detected.

Example. In this example we develop an interface circuit and an Arduino sketch as a smoke detector. We use the MQ–4 sensor as the active element. The manufacturer's data sheet recommends a load resistance of 20 kOhms. A 100 kOhm potentiometer, set for 20 kOhms, serves as the load resistor as shown in Fig. 3.26. In clear air Vout is 2.0 V. In the presence of smoke Vout approaches 4.0 V. The sensitivity level of the circuit may be set by adjusting the potentiometer.

The following sketch senses Vout and activates a buzzer and LED when smoke is detected at a set sensitivity level. The threshold value for activation is set for Vout at 2.5 VDC.

```
//******************************************************************
//MQ_test:  This sketch tests the operation of an MQ-4 sensor.
//          When the output voltage from the sensing circuit exceeds
//          2.5 VDC an active low buzzer is activated and an LED is
//          flashed until the smoke clears.
//******************************************************************

#define buzzer_pin    2            //digital pin - active low buzzer
#define LED_pin       3            //digital pin - LED connection
#define MQ_sensor_pin A0           //analog pin  - MQ sensor

int MQ_sensor_value;               //variable for MQ sensor value

void setup()
  {
  pinMode(buzzer_pin, OUTPUT);     //configure pin 2 for digital output
  pinMode(LED_pin, OUTPUT);        //configure pin 3 for digital output
  digitalWrite(LED_pin,    LOW);   //turn LED off
  digitalWrite(buzzer_pin, HIGH);  //turn buzzer off - active low
  }

void loop()
```

---

[1] Data sheets for many of the MQ sensors are provided at www.mysensors.org.

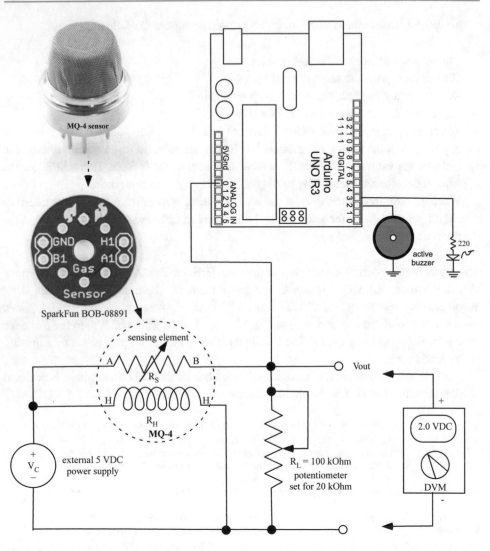

Fig. 3.26 MQ sensor test circuit. Sensor images courtesy of Sparkfun Electronics, (CC BY–NC–SA) (www.sparkfun.com)

```
  {
                                    //read analog output from MQ sensor
MQ_sensor_value = analogRead(MQ_sensor_pin);

if(MQ_sensor_value > 512)      //0 to 1023 maps to 0 to 5 VDC
  {
  digitalWrite(buzzer_pin, LOW);   //turn buzzer on - active low
  digitalWrite(LED_pin,    HIGH); //turn LED on - flash LED
  delay(100);                      //delay 100 ms
```

```
    digitalWrite(LED_pin,    LOW);   //turn LED off
    delay(100);                      //delay 100 ms
    digitalWrite(LED_pin,    HIGH);  //turn LED on
    delay(100);                      //delay 100 ms
    digitalWrite(LED_pin,    LOW);   //turn LED off
    }
  else
    {
    digitalWrite(buzzer_pin, HIGH);  //turn buzzer off
    digitalWrite(LED_pin,    LOW);   //turn LED off
    }
}

//**********************************************************************
```

The MQ series of sensors may be calibrated to provide a reading in gas parts per million, or "PPM." The interested reader is referred to procedures in the appropriate data sheet (Egironi 2018) [7].

3.10   Application: Dagu Rover 5–2, Two Motors, Two Encoders, with Wheels

Recall from Chap. 1 the Dagu Rover 5 robots are available in three variants. In this application, we use the Dagu Rover 5–2 robot for control by an Arduino UNO R3 as a maze navigating robot. The ROV5–2 is a tracked robot equipped with two motors and two motor encoders. We replace the robot tracks with Dagu 28 mm wheels equipped with 4 mm hubs as shown in Fig. 1.23. The robot is controlled by two 7.2 VDC motors which independently drive a left and right wheel. Two wheels are equipped with quadrature encoders. It is important to note that an interrupt input is required for each quadrature encoder equipped wheel. In this application we use UNO R3 interrupts INT0 and INT1. Please review the Application section of Chap. 1 before proceeding. We only highlight the addition of the quadrature encoders here to count wheel interrupts to monitor robot turns. The modified circuit diagram with the quadrature encoders is shown in Fig. 3.27.

3.10.1   Microcontroller Code–Arduino UNO

Provided here is the modified sketch incorporating the quadrature encoders to count wheel interrupts and hence motor turns. Due to space limitations, we have provided representative code to move forward, turn left, and turn right.

In this example we use two interrupts (INT0 and INT1) to monitor wheel odometry. To prevent the interrupts from conflicting with one another, note the use of "detachInterrupt" and "attachInterrupt" at appropriate times to turn off and on an interrupt.

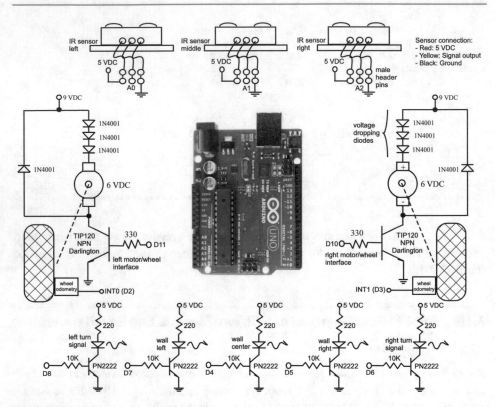

Fig. 3.27 Robot circuit diagram with quadrature encoders. UNO R3 illustration used with permission of the Arduino Team (CC BY–NC–SA) (www.arduino.cc) [3]

```
//***********************************************************************
//Dagu 5-2 Maze Following Robot
//
//Turns are rendered using wheel odometry.  Odometry counts are
//tracked using interrupts
//
//Left  wheel: INT0 - tracks left wheel odometry interrupts
//Right wheel: INT1 - tracks right wheel odometry interrupts
//***********************************************************************

#define left_IR_sensor    A0          //analog pin - left IR sensor
#define center_IR_sensor  A1          //analog pin - center IR sensor
#define right_IR_sensor   A2          //analog pin - right IR sensor

                                      //LED indicators - wall indicators
#define wall_left         7           //digital pin - wall_left
#define wall_center       4           //digital pin - wall_center
#define wall_right        5           //digital pin - wall_right

                                      //LED indicators - turn signals
#define left_turn_signal  8           //digital pin - left_turn_signal
```

```
#define right_turn_signal 6              //digital pin - right_turn_signal


                                         //motor outputs
#define left_motor       11              //digital pin - left_motor
#define right_motor      10              //digital pin - right_motor


                                         //interrupt inputs
#define l_int            2               //digital pin - left_interrupt
#define r_int            3               //digital pin - right_interrupt

int l_sensor;                            //variable for left IR sensor
int c_sensor;                            //variable for center IR sensor
int r_sensor;                            //variable for right IR sensor
char turn_signals;                       //required turn signals
volatile int desired_l_wheel_int, actual_l_wheel_int;
volatile int desired_r_wheel_int, actual_r_wheel_int;
int last_action;                         //0: off, 1: on
int troubleshoot = 1;

void setup()
{
if(troubleshoot) Serial.begin(9600);
                                         //LED indicators - wall indicators
pinMode(wall_left,    OUTPUT);           //configure pin 1 for digital output
pinMode(wall_center,  OUTPUT);           //configure pin 2 for digital output
pinMode(wall_right,   OUTPUT);           //configure pin 3 for digital output


                                         //LED indicators - turn signals
pinMode(left_turn_signal,OUTPUT);        //configure pin 0 for digital output
pinMode(right_turn_signal,OUTPUT);       //configure pin 4 for digital output


                                         //motor outputs - PWM
pinMode(left_motor,   OUTPUT);           //config pin 11 for digital output
pinMode(right_motor,  OUTPUT);           //config pin 10 for digital output
}

void loop()
{
                                         //read analog output from IR sensors
                                         //detect obstructions
l_sensor    = analogRead(left_IR_sensor);
c_sensor    = analogRead(center_IR_sensor);
r_sensor    = analogRead(right_IR_sensor);
                                         //reset wall LEDs
digitalWrite(wall_left,    LOW);         //turn LED off
digitalWrite(wall_center,  LOW);         //turn LED off
digitalWrite(wall_right,   LOW);         //turn LED off

//robot action table row 0
//no walls detected - continue forward
if((l_sensor < 512)&&(c_sensor < 512)&&(r_sensor < 512))
  {
  if (troubleshoot) Serial.println("row 0");
  //illuminate LEDs indicating walls detected
  digitalWrite(wall_left,    LOW);       //turn LED off
```

```
    digitalWrite(wall_center, LOW);      //turn LED off
    digitalWrite(wall_right,  LOW);      //turn LED off

    //illuminate turn signal - none required moving forward
    turn_signals = 'n';                  //turn signal action
    last_action = 0;                     //0: off, 1: on

    //activate motors - left on, right on
    analogWrite(left_motor,  128);       //0 (off) to
                                         //255 (full speed)
    analogWrite(right_motor, 128);       //0 (off) to
                                         //255 (full speed)
    //reset actual counter, set desired counter
    actual_l_wheel_int = 0;              //actual int count reset
    actual_r_wheel_int = 0;              //actual int count reset

    desired_l_wheel_int = 150;            //set int for hard left
    desired_r_wheel_int = 150;            //set int for hard left
                                          //monitor either wheel
                                          //wait for interrupts

    //Enable the interrupt that will be counted, disable the other
    detachInterrupt(digitalPinToInterrupt(l_int));
    attachInterrupt(digitalPinToInterrupt(r_int), int1_ISR_r_wheel_cnt, RISING);
    while (actual_r_wheel_int < desired_r_wheel_int);
    detachInterrupt(digitalPinToInterrupt(r_int));
    if (troubleshoot) Serial.println(actual_l_wheel_int);
    if (troubleshoot) Serial.println(actual_r_wheel_int);

    analogWrite(left_motor,  0);         //turn motors off
    analogWrite(right_motor, 0);
                                         //turn signals off
    digitalWrite(left_turn_signal,  LOW);//turn LED off
    digitalWrite(right_turn_signal, LOW);//turn LED off

    }

//robot action table row 1
//wall on right detetected - continue forward
else if((l_sensor < 512)&&(c_sensor < 512)&&(r_sensor > 512))
    {
      :
    Use code similar to row 0
    }

//robot action table row 2
//wall in front detected - turn right
else if((l_sensor < 512)&&(c_sensor > 512)&&(r_sensor < 512))
    {
    if (troubleshoot) Serial.println("row 2");
    //illuminate LEDs indicating walls detected
    digitalWrite(wall_left,   LOW);      //turn LED off
    digitalWrite(wall_center, HIGH);     //turn LED on
    digitalWrite(wall_right,  LOW);      //turn LED off
```

```
//illuminate turn signal - turning right
turn_signals = 'r';                    //turn signal action
last_action = 0;                       //0: off, 1: on

//activate motors - left on, right off
analogWrite(left_motor,  128);         //0 (off) to
                                       //255 (full speed)
analogWrite(right_motor, 0);           //0 (off) to
                                       //255 (full speed)
//reset actual counter, set desired counter
actual_l_wheel_int = 0;                //actual int count reset
actual_r_wheel_int = 0;                //actual int count reset

desired_l_wheel_int = 150;             //set int for hard left
desired_r_wheel_int = 150;             //set int for hard left
                                       //monitor either wheel
                                       //wait for interrupts
//Enable the interrupt that will be counted, disable the other
attachInterrupt(digitalPinToInterrupt(l_int), int0_ISR_l_wheel_cnt, RISING);
detachInterrupt(digitalPinToInterrupt(r_int));
while (actual_l_wheel_int < desired_l_wheel_int);
detachInterrupt(digitalPinToInterrupt(l_int));
if (troubleshoot) Serial.println(actual_l_wheel_int);
if (troubleshoot) Serial.println(actual_r_wheel_int);

analogWrite(left_motor,  0);           //turn motors off
analogWrite(right_motor, 0);
                                       //turn signals off
digitalWrite(left_turn_signal,  LOW);//turn LED off
digitalWrite(right_turn_signal, LOW);//turn LED off
}

//robot action table row 3
//wall in front and right detected - turn left
else if((l_sensor < 512)&&(c_sensor > 512)&&(r_sensor > 512))
{
if (troubleshoot) Serial.println("row 3");
//illuminate LEDs indicating walls detected
digitalWrite(wall_left,   LOW);       //turn LED off
digitalWrite(wall_center, HIGH);      //turn LED on
digitalWrite(wall_right,  HIGH);      //turn LED on

//illuminate turn signal - none required moving forward
turn_signals = 'l';                    //turn signal action
last_action = 0;                       //0: off, 1: on

//activate motors - left off, right on
analogWrite(left_motor,  0);           //0 (off) to
                                       //255 (full speed)
analogWrite(right_motor, 128);         //0 (off) to
                                       //255 (full speed)
//reset actual counter, set desired counter
actual_l_wheel_int = 0;                //actual int count reset
actual_r_wheel_int = 0;                //actual int count reset
```

```
  desired_l_wheel_int = 150;               //set int for hard left
  desired_r_wheel_int = 150;               //set int for hard left
                                           //monitor either wheel
                                           //wait for interrupts
  detachInterrupt(digitalPinToInterrupt(l_int));
  attachInterrupt(digitalPinToInterrupt(r_int), int1_ISR_r_wheel_cnt, RISING);
  while (actual_r_wheel_int < desired_r_wheel_int);
  detachInterrupt(digitalPinToInterrupt(r_int));
  if (troubleshoot) Serial.println(actual_l_wheel_int);
  if (troubleshoot) Serial.println(actual_r_wheel_int);

  analogWrite(left_motor,  0);             //turn motors off
  analogWrite(right_motor, 0);
                                           //turn signals off
  digitalWrite(left_turn_signal,  LOW);//turn LED off
  digitalWrite(right_turn_signal, LOW);//turn LED off
  }

//robot action table row 4
//wall on left - continue forward
else if((l_sensor > 512)&&(c_sensor < 512)&&(r_sensor < 512))
  {
     :
  Use code similar to row 0
  }

//robot action table row 5
//wall on left and right - continue forward
else if((l_sensor > 512)&&(c_sensor < 512)&&(r_sensor > 512))
  {
     :
  Use code similar to row 0
  }

//robot action table row 6
//wall on left and in front detected - turn right
else if((l_sensor > 512)&&(c_sensor > 512)&&(r_sensor < 512))
  {
     :
  Use code similar to row 2
  }

//robot action table row 7
//wall on left, front, and right detected - turn right
else if((l_sensor > 512)&&(c_sensor > 512)&&(r_sensor > 512))
  {
     :
  Use code similar to row 2
  }
}


//****************************************************************
//int0_ISR_l_wheel_cnt: interrupt service routine for INT0
//****************************************************************
```

```
void int0_ISR_1_wheel_cnt(void)
{
//increment actual interrupt count
actual_1_wheel_int++;

//if (troubleshoot) Serial.println("INT0 interrupt");
//if (troubleshoot) Serial.println(actual_1_wheel_int);


//flash turn signals
switch(turn_signals)
  {
  case 'n': //turn signals off
          digitalWrite(left_turn_signal,  LOW);    //turn LED off
          digitalWrite(right_turn_signal, LOW);    //turn LED off
          break;

   case 'l': //toggle left turn signal
          if (last_action == 0) //off
             {
             digitalWrite(left_turn_signal,HIGH);   //turn LED on
             last_action = 1;
             }
          else
             {
             digitalWrite(left_turn_signal,LOW);    //turn LED off
             last_action = 0;
             }
          digitalWrite(right_turn_signal, LOW);    //turn LED off
          break;

   case 'r': //toggle right turn signal
          if (last_action == 0) //off
             {
             digitalWrite(right_turn_signal,HIGH); //turn LED on
             last_action = 1;
             }
          else
             {
             digitalWrite(right_turn_signal,LOW); //turn LED off
             last_action = 0;
             }
          digitalWrite(left_turn_signal,  LOW);   //turn LED off
          break;

   default: ;
   }
}

//****************************************************************
//int1_ISR_r_wheel_cnt: interrupt service routine for INT1
//****************************************************************

void int1_ISR_r_wheel_cnt(void)
{
```

```
//increment actual interrupt count
actual_r_wheel_int++;

//if (troubleshoot) Serial.println("INT1 interrupt");
//if (troubleshoot) Serial.println(actual_r_wheel_int);

//flash turn signals
switch(turn_signals)
  {
  case 'n': //turn signals off
            digitalWrite(left_turn_signal,  LOW);   //turn LED off
            digitalWrite(right_turn_signal, LOW);   //turn LED off
            break;

  case 'l': //toggle left turn signal
            if (last_action == 0) //off
               {
               digitalWrite(left_turn_signal,HIGH);  //turn LED on
               last_action = 1;
               }
            else
               {
               digitalWrite(left_turn_signal,LOW);   //turn LED off
               last_action = 0;
               }
            digitalWrite(right_turn_signal, LOW);   //turn LED off
            break;

  case 'r': //toggle right turn signal
            if (last_action == 0) //off
               {
               digitalWrite(right_turn_signal,HIGH); //turn LED on
               last_action = 1;
               }
            else
               {
               digitalWrite(right_turn_signal,LOW); //turn LED off
               last_action = 0;
               }
            digitalWrite(left_turn_signal,  LOW);   //turn LED off
            break;

  default: ;
  }
}

//****************************************************************
```

3.11 Summary

The goal of this chapter was to investigate common concepts for robot applications. In a given application, all concepts may or may not be employed. The concepts are equally applicable from small, simple robots to complex systems. We discussed concepts related to robot location, steering, vision and obstacle avoidance, odometry, status monitoring, control, and autonomous versus remote control. From a robot's point of view, the concepts in this chapter provided answers to questions such as:

- Where am I on the Earth's surface?
- Are there obstacles near by I need to avoid?
- How will I move about?
- What is my orientation and movement within a local X, Y, Z coordinate system?
- What direction am I heading relative to a compass setting?
- What are the characteristics of the environment around me?

3.12 Problems

1. What is the accuracy GPS?
2. Can GPS be used indoors? Explain.
3. Provide a table of GPS related NME messages.
4. What is the difference between GMT and UTC time?
5. What is the difference between latitude and longitude?
6. What is the significance of the Prime Meridian?
7. What features due Mecanum wheels provide to a robot? Explain.
8. How may you control the speed, direction, and speed and direction of a DC motor?
9. Describe the difference between single channel and dual channel odometry.
10. In the dual channel odometry example, why was the output from each channel EXORed together?
11. What is the wavelength of IR light?
12. The profile of the Sharp GP2Y0A41SK0F IR sensor provides the same voltage for two different ranges. How may this ambiguity be removed in a robot application? Hint: Consider the physical placement of the sensor.
13. What is the sound frequency of the signal from an ultrasonic sensor? Can a human hear this? Why or why not? Explain.
14. What are the features of laser light?
15. What was the purpose of the LM 324 op amps in the pan and tilt application? Explain.

16. Compare and contrast the features of an alphanumeric LCD and a TFT display.
17. Construct a structure chart and an UML activity diagram for the algorithm used in the Application section.

References

1. L. Ada, *Adafruit 2.8" TFT Touch Shield V2–Capacitive or Resistive*, www.adafruit.com, June 2020.
2. L. Ada, *Adafruit Ultimate GPS Logger Shield*, www.adafruit.com, Dec 2020.
3. Arduino homepage, www.arduino.cc
4. P. Burgess, *Adafruit GFX Graphics Library*, www.adafruit.com, Nov 2020.
5. *Dagu Mini Pan and Tilt Kit with 2 Servos*, data sheet, Jameco #2157870, www.jameco.com.
6. O. Diegel, A. Badve, G. Brighht, J. Potgieter, and S. Tlale, *Improved Mecanum Wheel Design for Omni–directional Robots,* Proc. 2002 Australasian Conference on Robotics and Automation, Auckland, November 2002.
7. D. Egironi, *Presenting MQ sensors: low–cost gas and pollution detectors,* Open Electronics Source Electronic Projects, February 2018,
8. *FGPMMOPA6H GPS Standalone Module Data Sheet, Rev V0A*, GlobalTop Technology, Inc, 2011.
9. *Garmin Lidar Lite V3 Operation Manual and Technical Specifications*, Garmin International, Inc, Olathe, Kansas, Sep 2016.
10. *Global Positioning System Precise Positioning Service Performance Standard*, www.gps.gov, Feb 2007.
11. *Greenwich Mean Time,* www.greenwichmeantime.com
12. *How to Use a Quadrature Encoder*, www.dagu.com.
13. J.B. Kuipers, *Quaternions and Rotation Sequences–A Primer with Applications to Orbits, Aerospace, and Virtual Reality,* Princeton University Press, 1999.
14. G. McComb, *Ways to move your robot,* Servo Magazine, May 2014.
15. *LV–Max Sonar–EZ Series High Performance Sonar Range Finder*, MaxBotix Inc, www.maxbotix.com, 2015
16. *MQ–2 Semiconductor Sensor for Combustible Gas,* www.mysensors.org
17. *MQ–3 Gas Sensor,* Hanwei Electronics Co., ltd., www.hwsensor.com.
18. *MQ–4 Gas Sensor,* Hanwei Electronics Co., ltd., www.hwsensor.com.
19. *MQ–6 Flammable Gas Sensor,* Zhengzhou Winsen Electronics Technology Co., Ltd, www.winsensor.com
20. *MQ–7 Gas Sensor,* Hanwei Electronics Co., ltd., www.hwsensor.com.
21. *MQ–8 Flammable Gas Sensor,* Zhengzhou Winsen Electronics Technology Co., Ltd, www.winsensor.com.
22. *MQ–9 Semiconductor Sensor for CO/Combustible Gas,* Hanwei Electronics Co., ltd., www.hwsensor.com.
23. *MQ–131 Gas Sensor,* www.mysensors.org.
24. *MQ–135 Gas Sensor,* www.mysensors.org.
25. *MQ–183 Gas Sensor,* Hanwei Electronics Co., ltd., www.hwsensor.com.
26. *MQ–214 Gas Sensor,* Hanwei Electronics Co., ltd., www.hwsensor.com.

27. *Sharp GP2Y0A41SK0F Distance Measuring Sensor Unit,* Sheet no. OP113008EN.
28. B. Siepert, *Adafruit 9–DOF Orientation IMU Fusion Breakout–BNO085,* www.adafruit.com, March 2021.
29. SparkFun Electronics, 6175 Longbow Drive, Suite 200, Boulder, CO 80301 (www.sparkfun. com)

Motor Control and Actuators

4

Objectives: After reading this chapter, the reader should be able to

- Summarize and provide definitions for DC motor parameters and ratings;
- Describe the need for and design a motor interface for different types of DC motors;
- Describe how to control the direction and speed of a DC motor;
- Discuss the requirement for an optical based motor interface;
- Summarize different types of motors used by robots and provide sample applications; and
- Equip and control a robot with different types of DC motors.

4.1 Overview Concepts

In this chapter we investigate different types of DC motors used onboard robots. The motors may be used for locomotion, sensor positioning and scanning, or actuator positioning. For each type of motor we provide a basic theory of operation, microcontroller interface techniques, and example applications.[1]

Some of the different types of motors available for robot applications are shown in Fig. 4.1.

- **DC motor:** A DC motor has a positive and negative terminal. When a DC power supply of suitable voltage and current rating is applied to the motor it will rotate. If the polarity of the supply is switched with reference to the motor terminals, the motor will rotate in the opposite direction. The speed of the motor is roughly proportional to the applied voltage up to the rated voltage of the motor.

---

[1] Portions of this chapter appeared in "Arduino I Getting Started," S. Barrett, Morgan and Claypool Publishers, 2020. The information is provided for completeness.

© The Author(s), under exclusive license to Springer Nature Switzerland AG 2022
T. Kerr and S. Barrett, *Arduino IV: DIY Robots*, Synthesis Lectures on Digital Circuits
& Systems, https://doi.org/10.1007/978-3-031-11209-6_4

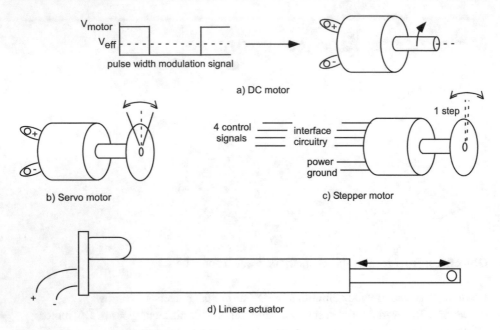

Fig. 4.1 Motor types

- **Servo motor:** A servo motor provides a precision angular rotation for an applied pulse width modulation (PWM) duty cycle. As the duty cycle of the applied signal is varied, the angular displacement of the motor also varies. This allows the motor to be used in applications to change mechanical positions such as the steering angle of a wheel, position or scan a sensor, or as the main drive motor for a small robot.
- **Stepper motor:** A stepper motor, as its name implies, provides an incremental step change in rotation (typically 2.5° per step) for a step change in control signal sequence. The motor is typically controlled by a two or four wire interface. For the four wire stepper motor, the microcontroller provides a four bit control sequence to rotate the motor clockwise. To turn the motor counterclockwise, the control sequence is reversed. The low power microcontroller control signals are interfaced to the motor via MOSFETs or power transistors to provide for the proper voltage and current requirements of the pulse sequence. The stepper motor may be used to position or scan robot sensors.
- **Linear actuator** The linear actuator is actually a rotating DC motor equipped with gears to translate rotational motion to linear motion. The linear actuator is used when repeatable linear motion, both push and pull, is required. They may be used in robots for steering or for sensor placement.

With this brief overview of motors complete, let's take a closer look at each motor type.

4.2   DC Motor

A direct current or DC motor is typically used in robot applications as the main source of locomotion. The power source for the DC motor is usually an onboard battery supply carried by the robot. We discuss the choice of battery supplies in the next chapter.

A DC motor has a positive and negative terminal. When a DC power supply of suitable voltage and current rating is applied to the motor it will mechanically rotate. If the polarity of the supply is switched with reference to the motor terminals, the motor will rotate in the opposite direction. The speed of the motor is roughly proportional to the applied voltage up to the rated voltage of the motor.

4.2.1   DC Motor Ratings

A DC motor is rated using the following common parameters. The requirements of the robot are used to select an appropriate motor [6, 8].

- **voltage:** The maximum operating voltage of a motor is specified in DC volts. At this voltage the motor rotates at its maximum speed specified in revolutions per minute or RPM.
- **current:** When rotating, a motor will draw current from the DC supply. The current, measured in amperes or amps, drawn depends on the load the motor is experiencing. DC motors have the following currents specified: starting current, no load operating current, and stall current. When a motor interface is designed it must withstand these different current values.
  - ∗ **start current:** The start current is a surge current that occurs when a motor first starts. The surge is due to the motor overcoming mechanical inertia.
  - ∗ **no load current:** The no load operating current is the value of current drawn when the motor is supplied its rated voltage and is not under a mechanical load.
  - ∗ **stall current:** The stall current is the current drawn from the supply when the motor is stalled.
- **speed:** The rotational speed of a motor is specified in revolutions per minute or RPM.
- **torque:** Torque is the angular force delivered by the motor at a radius from the motor shaft. It is measured in MKS units of Newton–meters.
- **speed versus torque:** DC motors have several different types of configurations: shunt, compound, and series. Generally speaking, as the motor shaft speed is decreased (e.g. via gears), the motor torque is increased.
- **efficiency:** A DC motor converts DC electrical power into mechanical power. Ideally, we desire all of the electrical power to convert to mechanical power representing 100 percent efficiency. Efficiency is defined as the ratio of mechanical power to electrical power.

- **gears:** Gears are used to reduce the shaft speed of a DC motor while increasing motor torque. Many DC motors are equipped with a gearbox for this purpose.

4.2.2   Unidirectional DC Motor Control

In this section we describe the interface between the low power Arduino microcontroller and a DC motor. In "Arduino I Getting Started," Chap. 3 "Arduino Power and Interfacing," the Arduino UNO R3 CMOS operating parameters are provided. The output voltage for a logic high is found to be 4.2 V with a maximum current of 0.8 mA. At heavier current loads the output voltage is pulled down to lower values.[2]

These voltage and current values are not sufficient to directly drive any type of motor. Therefore, an interface is required to boost the voltage and current to values consistent with a given motor specification. Two common methods of interface include using a Darlington transistor configuration or a power metal–oxide–semiconductor field–effect transistor (MOSFET).

4.2.2.1 NPN Darlington Transistor

A Darlington configuration consists of two transistors configured as shown in Fig. 4.2a. The emitter of the first transistor is connected to the base of the second transistor. This configuration provides for high current gain. When used as a motor interface, the low output current control signal from the Arduino is boosted to a current value suitable to drive a motor as shown in (b). A series of silicon diodes (1N4001), each with a voltage drop of 0.7 VDC, is used to drop the power supply voltage down to the motor supply voltage rating. Additionally, a reverse biased protection diode is placed across the diode string and motor (Boylestad and Nashelsky 2002, Sedra and Smith 2004).

Example: The Dagu Magician is a popular, low–cost, two platform robot. The robot is equipped with two drive motors rated at 6 VDC with a maximum current rating of 250 mA and a stall current of 1 amp. In this application, the robot motors are powered from an external 9 VDC power supply via an umbilical cable.

We assume an Arduino UNO R3 output voltage of 4.5 VDC with a maximum output current of approximately 8.5 mA. These values are obtained from the output characteristics of the ATmega328 provided in "Arduino I: Getting Started."[3] With this information, the base resistor value for the Darlington transistor is calculated to be approximately 360 Ohms.

---

[2] S.F. Barrett, "Arduino I Getting Started," Morgan & Claypool Publishers, 2020.

[3] The Microchip ATmega328 is the host microcontroller onboard the Arduino UNO R3.

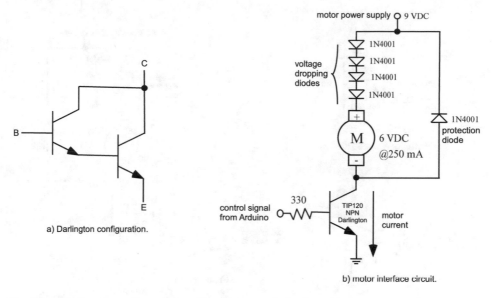

Fig. 4.2 **a** Darlington configuration and **b** motor control circuit

$$-V_{OH} + (I_B \times R_B) + (2 \times V_{BE}) = 0$$

$$-4.5 + (0.0085 \times R_B) + (2 \times 0.7) = 0$$

360 Ohms

We use a close standard resistor value of 330 Ohms as shown in Fig. 4.2b. The NPN Darlington transistor (TIP 120) boosts the base current to a collector current value suitable to supply the motor. A separate interface circuit is required for each motor.

4.2.2.2 High Power MOSFET

An N–channel enhancement power MOSFET may also be used to switch a high current load on and off using a low voltage control signal from a microcontroller as shown in Fig. 4.3a. The gate is electrically isolated from the high current capacity channel between the drain and the source. When a voltage is applied to the gate, it attracts charge carriers to build up or enhance the current carrying capacity of the drain to source channel (Boylestad and Nashelsky 2002).

When the control signal on the MOSFET gate is logic high, the load current flows from drain to source. When the control signal applied to the gate is logic low, no load current flows. Thus, the high power load is turned on and off by the low power control signal from the microcontroller.

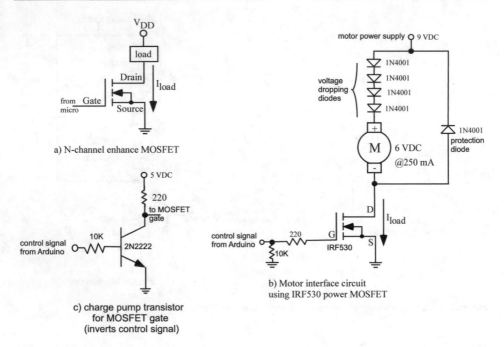

Fig. 4.3 MOSFET circuits

Often the MOSFET is used to control a high power motor load. For an inductive load such as a motor, a reversed biased protection diode is placed across the load. It is important to note that the load supply voltage and the microcontroller supply voltage do not have to be at the same value.

Example: In this example we again use the Dagu Magician robot equipped with two drive motors rated at 6 VDC with a maximum current rating of 250 mA and a stall current of 1 amp. In this application, the robot motors are powered from an external 9 VDC power supply via an umbilical cable. The interface circuit using an IRF530 power MOSFET is shown in Fig. 4.3b.

For higher power loads it is helpful to equip the MOSFET gate with a charge pump circuit to ensure the on/off switch speed of the MOSFET. A simple charge pump circuit consisting of a 2N2222 NPN transistor is shown in Fig. 4.3c. We explore dedicated charge pump integrated circuits in an upcoming section. A separate interface circuit is required for each motor.

4.2.2.3 Optical Interface for Noise Isolation
A motor is a notorious source of electrical noise. To isolate the microcontroller control signal from the motor noise an optical isolator may be used as an interface as shown in Fig. 4.4a.

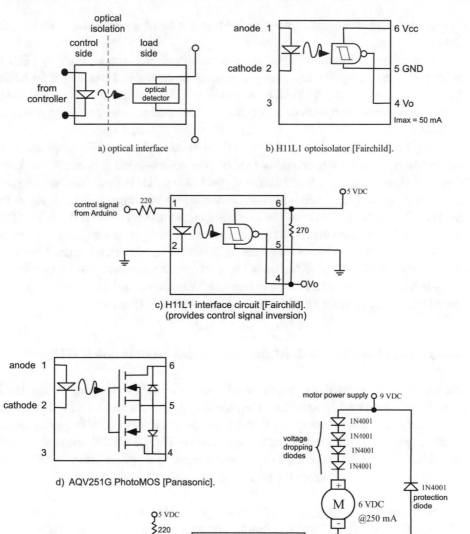

a) optical interface

b) H11L1 optoisolator [Fairchild].

c) H11L1 interface circuit [Fairchild].
(provides control signal inversion)

d) AQV251G PhotoMOS [Panasonic].

d) AQV251G PhotoMOS
(noninverting control signal)

Fig. 4.4 Optical isolation

The link between the microcontroller control signal to the high power load is via an optical link.

The microcontroller control signal is converted to an optical version using an LED. The now optical control signal bridges the gap between the control and load side of the isolator. The optical detector on the load side converts the optical signal back to an electrical version. To ensure noise isolation between the control and load side, separate power supplies should be used.

The H11L1 optical isolator is shown in Fig. 4.4b, c. An internal LED is turned on and off consistent with the microcontroller signal. The optical signal is converted back to an electrical signal using an optical Schmitt trigger. **Note:** the H11L1 provides an inversion of the control signal. That is, when the microcontroller signal is high, the output signal from the H11L1 is low and vice versa. The maximum output rating for the H11L1 is 16 VDC at 50 mA. The output from the H11L1 is then routed to a high power interface circuit such as the Darlington configuration or the power MOSFET previously discussed (Fairchild).

The AQV251G photoMOS shown in Fig. 4.4d combines optical isolation and the capability to drive a motor load as shown in Fig. 4.4e. With a maximum rating of 30 V at 3.5 A, the AQV251G can drive the 6 VDC, 250 mA motor directly (Panasonic).

4.2.3   DC Motor Speed Control–Pulse Width Modulation (PWM)

As previously mentioned, DC motor speed may be varied by changing the applied DC voltage. However, PWM control signal techniques may be combined with a motor interface to precisely control the motor speed. With a PWM control signal, a fixed frequency and duty cycle is provided to the motor interface. As shown in Fig. 4.5 the duty cycle of the PWM signal will also be the percentage of the motor supply voltage, or effective DC voltage, applied to the motor and hence the percentage of rated full speed at which the motor will rotate.

The Arduino UNO R3 is equipped with several PWM channels. These are accessible on pins 3, 5, 6, 9, 10, and 11. The PWM baseline frequency on pins 5 and 6 is 980 Hz, while on the remaining pins (3, 9, 10, and 11), the baseline PWM frequency is 490 Hz (arduino.cc).

4.2.4   Bidirectional Motor Control with an H–bridge

For a DC motor to operate in both the clockwise and counter clockwise direction, the polarity of the DC motor supplied must be changed. To operate the motor in the forward direction, the positive battery terminal must be connected to the positive motor terminal while the negative battery terminal must be provided to the negative motor terminal. To reverse the motor direction the motor supply polarity must be reversed. An H–bridge is a circuit employed to perform this polarity switch. The H–bridge circuit consists of four

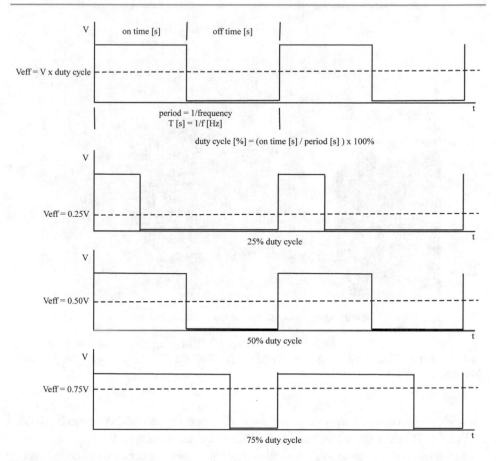

Fig. 4.5 Pulse width modulation

electronic switches as shown in Fig. 4.7. For forward motor direction switches 1 and 4 are closed; whereas, for reverse direction switches 2 and 3 are closed.

4.2.4.1 Integrated Circuit H–bridges

Texas Instruments (TI) provides a self–contained H–bridge motor controller integrated circuit, the DRV8829. Within the DRV8829 package is a single H–bridge driver. The driver may control DC loads with supply voltages from 8 to 45 VDC with a peak current rating of 5 A. The single H–bridge driver may be used to control a DC motor or one winding of a bipolar stepper motor (DRV8829).

Example: MIKROE–1526 DC MOTOR click. MikroElectronica (www.mikroe.com) manufactures a number of motor interface products including the MIKROE–1526 DC MOTOR click motor driver board. The board features the Texas Instruments DRV8833RTY

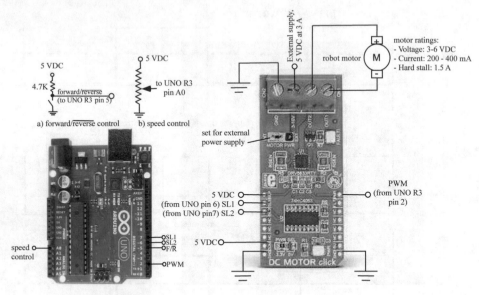

c) Arduino UNO R3 interface to MIKROE-1526 DC MOTOR click

Fig. 4.6 MIKROE-1526 DC MOTOR click. UNO R3 illustration used with permission of the Arduino Team (CC BY–NC–SA) (www.arduino.cc). MIKROE illustration used with permission (www.mikroe.com)

H–bridge motor driver. A test circuit to control a DC motor's speed and direction is provided in Fig. 4.6. We use a DC motor rated at 6 VDC with a stall current of 1.5 A.

In the test circuit, a tact switch is used to determine motor direction and a potentiometer for motor speed control. These two inputs are read by the Arduino program and proper control signals are issued to the MIKROE–1526 (SL1, SL2, and PWM) for motor speed and direction. The nSLP pin on the MIKROE–1526 must be logic high to enable the device. For the test circuit, the nSLP pin is tied to Vcc (5 VDC) (www.mikroe.com).

```
//****************************************************************
//MIKROE-1526 DC Motor click
//Sketch demonstrates operation of the MIKROE-1526 DC Motor click
//
//The circuit:
//  - Motor speed control potentiometer.
//    Potentiometer connected to analog pin A0.  The center wiper
//    pin of the potentiometer goes to the analog pin.  The side
//    pins of the potentiometer go to 5 VDC and ground.
//  - Motor forward/reverse control.
//    Tact switch connected to UNO R3 pin 5.
//  - MIKROE-1526 connections:
//    -- Select 1 (SL1) to UNO R3 pin 6
//    -- Select 2 (SL2) to UNO R3 pin 7
```

```
//     -- PWM to UNO R3 pin 2
//     -- nSLEEP to Vcc (5 VDC) to enable device
//
//This example code is in the public domain.
//*****************************************************************

int analog_in = A0;                 //analog input (0 to 1023)
int analog_out = 2;                 //analog output (0 to 255)
int forward_reverse = 5;            //direction control
int select1 = 6;                    //motor direction control
int select2 = 7;                    //SL1, SL2
int speed_value;                    //potentiometer input value
int switch_value;
int output_value;

void setup()
{
pinMode(forward_reverse, INPUT);
pinMode(select1, OUTPUT);
pinMode(select2, OUTPUT);
}

void loop()
{
//Deteremine motor direction
switch_value = digitalRead(forward_reverse);
if(switch_value == HIGH)     //forward direction
  {
  digitalWrite(select1, LOW);
  digitalWrite(select2, LOW);
  }
else                         //reverse direction
  {
  digitalWrite(select1, LOW);
  digitalWrite(select2, HIGH);
  }

//read analog in value
speed_value = analogRead(analog_in);

//map to analog out range
output_value = map(speed_value, 0, 1023, 0, 255);

//update analog out value
analogWrite(analog_out, output_value);

delay(50);
}


//*****************************************************************
```

4.2.4.2 Discrete Component H–bridge

An H–bridge may be constructed from discrete components as shown in Fig. 4.7a, b. The ZTX451 and ZTX551 are NPN and PNP transistors with similar characteristics. They have a continuous collector current rating of 1 A with a peak current rating of 2 A. The 11DQ06

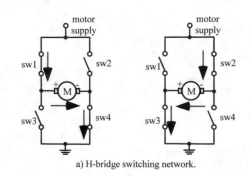

a) H-bridge switching network.

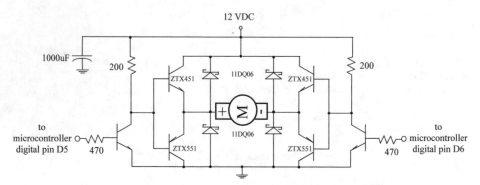

b) H-bridge to switch a 12 VDC, 1A load.

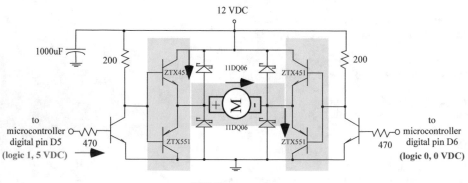

c) H-bridge current flow.

Fig. 4.7 H-bridge control circuit

are Schottky diodes. For driving higher power loads, the switching transistors are sized appropriately.

Figure 4.7c demonstrates how controls signals are used to control motor direction. A logic high is applied to one of the control signal inputs while a logic zero is applied to the other control signal input. Current flows to the motor as shown and the motor rotates. When the control signals are switched, the motor rotates in the opposite direction.

If PWM signals are used to drive the base of the transistors (from microcontroller pins D5 and D6), both motor speed and direction may be controlled by the circuit. The transistors used in the circuit must have a current rating sufficient to handle the current requirements of the motor during start and stall conditions.

In an upcoming section we discuss linear actuators and provide additional H–bridge control circuits.

4.3   Servo Motor Control

A servo motor provides a precision angular rotation for an applied pulse width modulation duty cycle. As the duty cycle of the applied signal is varied, the angular displacement of the motor also varies. The servo motor is used for a precise angular displacement. The displacement is related to the duty cycle of the applied control signal.

Example: Inexpensive Laser Light Show An inexpensive laser light show may be constructed from two servos.[4] In this example we use two Futaba 180 degree range servos (Parallax 900–00005, available from Jameco #283021) mounted as shown in Fig. 4.8. The X and Y control signals are provided by an Arduino processing board. The X and Y control signals are interfaced to the servos via LM324 operational amplifiers. The laser source is provided by an inexpensive laser pointer.

Sample code to drive the servos from an Arduino are provided on the Jameco website (www.jameco.com). Reference Jameco #283021.

```
//*************************************************************
//X-Y ramp
//*************************************************************

#include <Servo.h>      // Use Servo library, included with IDE

Servo myServo_x;        // Create Servo object to control the servo
Servo myServo_y;

void setup() {
  myServo_x.attach(9);    // Servo is connected to digital pin 9
  myServo_y.attach(10);   // Servo is connected to digital pin 10
```

---

[4] From "Arduino I: Getting Started, S.F. Barrett, Morgan and Claypool Publishers, 2020.

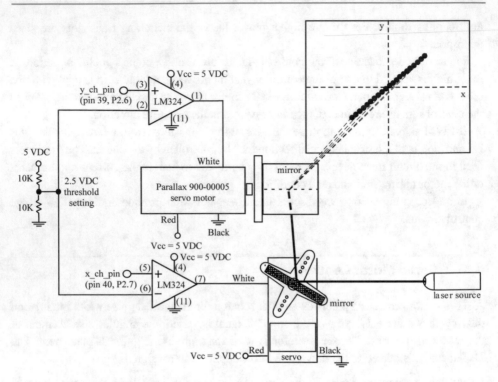

Fig. 4.8 Inexpensive laser light show

```
}

void loop() {
  int i = 0;
  for(i=0; i<=180; i++)
    {
    myServo_x.write(i);   // Rotate servo counter clockwise
    myServo_y.write(i);   // Rotate servo counter clockwise
    delay(20);            // Wait 2 seconds
    if(i==180)
      delay(5000);
    }
}
//*************************************************************
```

Example: In Chap. 3, servo motors were used to position an ultrasonic sensor and also a LIDAR system.

4.4   Stepper Motor Control

Stepper motors are used to provide a discrete angular displacement in response to a control signal step. There are a wide variety of stepper motors including bipolar and unipolar types with different configurations of motor coil wiring. Due to space limitations we only discuss the unipolar, 5 wire stepper motor. The internal coil configuration for this motor is shown in Fig. 4.9b.

Often, a wiring diagram is not available for the stepper motor. Based on the wiring configuration (Fig. 4.9b), one can determine the common line for both coils. It has a resistance that is one–half of all of the other coils. Once the common connection is found, one can connect the stepper motor into the interface circuit. By changing the other connections, one can determine the correct connections for the step sequence. To rotate the motor either clockwise or counterclockwise, a specific step sequence must be sent to the motor control wires as shown in Fig. 4.9b.

The microcontroller does not have sufficient capability to drive the motor directly. Therefore, an interface circuit is required as shown in Fig. 4.10. The speed of motor rotation is determined by how fast the control sequence is completed.

```
//***********************************************************
//stepper
//
//This example code is in the public domain.
//***********************************************************

//external switches
#define ext_sw1   9
#define ext_sw2   10

//stepper channels
#define stepper_ch1   5
#define stepper_ch2   6
#define stepper_ch3   7
#define stepper_ch4   8

int switch_value1, switch_value2;
int motor_speed = 1000;              //motor increment time in ms
int last_step = 1;
int next_step;

void setup()
{
//Screen
Serial.begin(9600);

//external switches
pinMode(ext_sw1,   INPUT);
pinMode(ext_sw2,   INPUT);
```

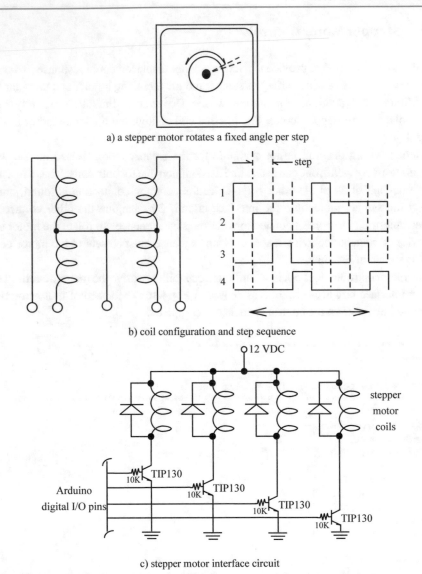

a) a stepper motor rotates a fixed angle per step

b) coil configuration and step sequence

c) stepper motor interface circuit

Fig. 4.9 Unipolar stepper motor interface circuit using TIP130 Darlington configured transistors

```
//stepper channel
pinMode(stepper_ch1, OUTPUT);
pinMode(stepper_ch2, OUTPUT);
pinMode(stepper_ch3, OUTPUT);
pinMode(stepper_ch4, OUTPUT);

}
```

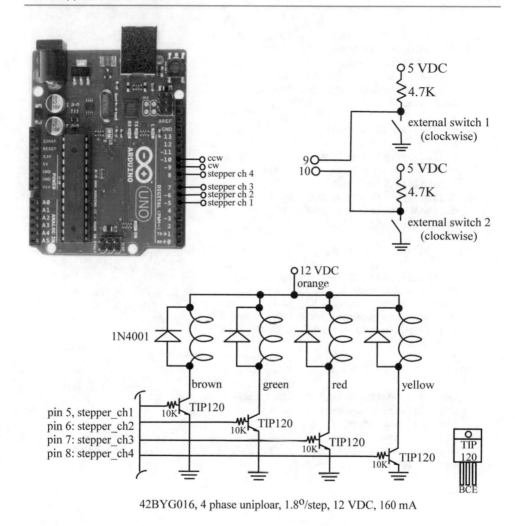

42BYG016, 4 phase uniploar, 1.8°/step, 12 VDC, 160 mA

Fig. 4.10 Unipolar stepper motor interface circuit. UNO R3 illustration used with permission of the Arduino Team (CC BY–NC–SA) (www.arduino.cc)

```
void loop()
{
switch_value1 = digitalRead(ext_sw1);
switch_value2 = digitalRead(ext_sw2);

if(switch_value1 == LOW)              //switch1 asserted
  {
  while(switch_value1 == LOW)         //clockwise
    {
    if(last_step == 1)
      {
```

```
      Serial.println("Switch 1: low, step 1");
      digitalWrite(stepper_ch1, HIGH);
      digitalWrite(stepper_ch2, LOW);
      digitalWrite(stepper_ch3, LOW);
      digitalWrite(stepper_ch4, LOW);
      next_step = 2;
      }
    else if(last_step == 2)
      {
      Serial.println("Switch 1: low, step 2");
      digitalWrite(stepper_ch1, LOW);
      digitalWrite(stepper_ch2, HIGH);
      digitalWrite(stepper_ch3, LOW);
      digitalWrite(stepper_ch4, LOW);
      next_step = 3;
      }
    else if(last_step == 3)
      {
      Serial.println("Switch 1: low, step 3");
      digitalWrite(stepper_ch1, LOW);
      digitalWrite(stepper_ch2, LOW);
      digitalWrite(stepper_ch3, HIGH);
      digitalWrite(stepper_ch4, LOW);
      next_step = 4;
      }
    else if(last_step == 4)
      {
      Serial.println("Switch 1: low, step 4");
      digitalWrite(stepper_ch1, LOW);
      digitalWrite(stepper_ch2, LOW);
      digitalWrite(stepper_ch3, LOW);
      digitalWrite(stepper_ch4, HIGH);
      next_step = 1;
      }
    else
      {
      ;
      }
    last_step = next_step;
    delay(motor_speed);
    switch_value1 = digitalRead(ext_sw1);
    }//end while
   }//end if

else if(switch_value2 == LOW)        //switch2 asserted
  {
  while(switch_value2 == LOW)        //counter clockwise
    {
    if(last_step == 1)
      {
      Serial.println("Switch 2: low, step 1");
      digitalWrite(stepper_ch1, HIGH);
      digitalWrite(stepper_ch2, LOW);
```

```
         digitalWrite(stepper_ch3, LOW);
         digitalWrite(stepper_ch4, LOW);
         next_step = 4;
         }
      else if(last_step == 2)
         {
         Serial.println("Switch 2: low, step 2");
         digitalWrite(stepper_ch1, LOW);
         digitalWrite(stepper_ch2, HIGH);
         digitalWrite(stepper_ch3, LOW);
         digitalWrite(stepper_ch4, LOW);
         next_step = 1;
         }
      else if(last_step == 3)
         {
         Serial.println("Switch 2: low, step 3");
         digitalWrite(stepper_ch1, LOW);
         digitalWrite(stepper_ch2, LOW);
         digitalWrite(stepper_ch3, HIGH);
         digitalWrite(stepper_ch4, LOW);
         next_step = 2;
         }
      else if(last_step == 4)
         {
         Serial.println("Switch 2: low, step 4");
         digitalWrite(stepper_ch1, LOW);
         digitalWrite(stepper_ch2, LOW);
         digitalWrite(stepper_ch3, LOW);
         digitalWrite(stepper_ch4, HIGH);
         next_step = 3;
         }
      else
         {
         ;
         }
      last_step = next_step;
      delay(motor_speed);
      switch_value2 = digitalRead(ext_sw2);
      }//end while
   }//end if

   else
      {
      digitalWrite(stepper_ch1, LOW);
      digitalWrite(stepper_ch2, LOW);
      digitalWrite(stepper_ch3, LOW);
      digitalWrite(stepper_ch4, LOW);
      }
}
//*************************************************************
```

Example. Adafruit (www.adafruit.com) manufactures a DC stepper motor breakout board (#3297) based on the Texas Instruments DRV8833RTY H–bridge motor driver. The board

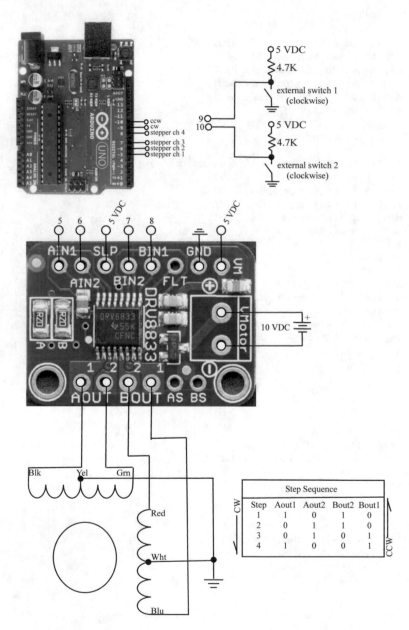

Fig. 4.11 Unipolar stepper motor with DRV8833 breakout board. DRV8833 breakout board image used courtesy of Adafruit (www.adafruit.com). UNO R3 illustration used with permission of the Arduino Team (CC BY–NC–SA) (www.arduino.cc)

can provide up to 1.2A per channel for motors from 2.7 to 10.8 VDC. In this example, we use the board to drive a Jameco (www.jameco.com) #238538 unipolar stepper motor rated at 12 VDC, 0.4A. We power the motor at 10 VDC. The interface between the Arduino UNO R3, the breakout board, and the stepper motor in shown in Fig. 4.11. Two external switches are used to select motor direction.

The code used in the previous stepper motor example may be modified with the step sequence required by the driver/motor combination.

4.5   Linear Motor/Actuator

A linear actuator is a specially designed motor that converts rotary to linear motion. The linear actuator is equipped with a mechanical rod that is extended when asserted in one direction and retracted when the polarity of assertion is reversed. An H–bridge may be used to control a linear actuator as shown in Fig. 4.12.

Example: In this example an Erfo micro linear actuator is controlled by an H–bridge. The linear actuator requires 12 VDC and has a 50 mm stroke length and a load capacity of 60N. It's no load speed is 15 mm/s. The current requirements are experimentally found to be 25 mA no load and 250 mA full load. For this example, we use the H–bridge configuration of Fig. 4.7b.

Example: In this example a high power linear actuator is used. We use a McMaster Carr 6509K83 linear actuator. It has a 6 inch stroke, requires a 12 VDC supply, and draws 6 A at a 25 pound load (McMaster–Carr). Due to the high power demand of the linear actuator, we use IRF3205PbF power MOSFETs as the H–bridge switching devices. Also, IR2104 half–bridge drivers are used as charge pumps for the MOSFETs as shoen in Fig. 4.12. In this example, a 13.6 VDC battery is used to power the circuit. An enable signal is used to assert the H–bridge while the two pulse width modulation channels are used to extend or retract the H–bridge.

Example: Several researchers have used surplus electric wheelchairs as robust robot platforms. Typically wheelchairs are equipped with two rugged DC drive motors. Benson designed an H–bridge for a Quikie P100 surplus wheelchair. The wheelchair hosted two Framco PM45/G31 DC motors. The motors were rated at 24 VDC at 7 A and 160 W. The motors were powered by two 12 VDC deep cycle batteries. Assuming a spike current of 35 amps, Benson developed an H–bridge for each motor using high power MOSFETS (Benson 2005).

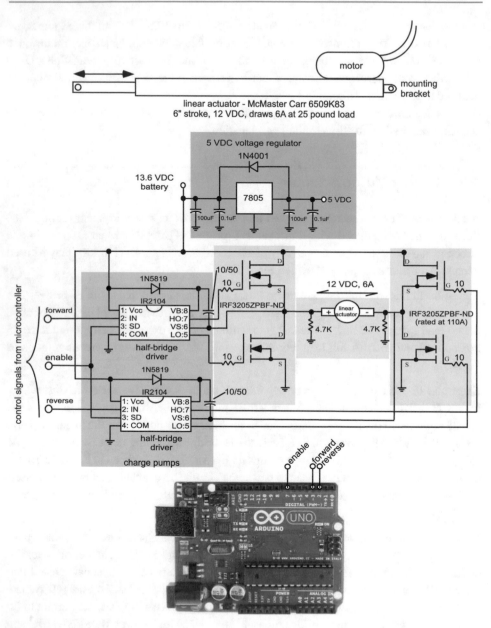

Fig. 4.12 Linear actuator control circuit (O'Berto). UNO R3 illustration used with permission of the Arduino Team (CC BY–NC–SA) (www.arduino.cc)

4.6   Application–4WD Platform with H–bridge

In the next two chapters we explore the full design of a 4WD platform based on the DF Robot 4WD mobile platform chassis with four wheels and motors (DF Robot #ROB0003).[5] In this section we develop the H–bridge control platform from the robot. For the overall project we use the Arduino Mega 2560 Rev 3 microcontroller board.

The DF Robot platform is equipped with four powered wheels. The motors are rated at 6 VDC and draw 160 mA at full speed. The locked rotor current is specified as 2.8 A. Mecanum wheels are available for the platform.[6]

The circuit diagram for the robot equipped with four H–bridges to independently control the direction and speed of each motor is provided in Fig. 4.13. The four robot motors are driven by eight separate PWM channels from Mega 2560 pins via four independent H–bridges. To operate a given motor in the forward direction, a PWM signal of desired duty cycle to provide desired motor speed is provided to the "fwd" signal of the H–bridge (e.g. lf_fwd). At the same time, a logic low is provided to the "rev" (e.g. lf_rev) input of the same H–bridge. Alternatively, to operate a motor in the reverse direction, a PWM signal of desired duty cycle to provide desired motor speed is provided to the "rev" signal of the H–bridge (e.g. lf_rev). Simultaneously, a logic low is provided to the "fwd" (e.g. lf_fwd) input of the same H–bridge.

Shown in Fig. 4.14 is a printed circuit board layout for four independent H–bridge control circuits. Note the use of TIP 31 (NPN) and TIP 32 (PNP) transistors. These transistors have a collector current rating of 3A (peak: 5A) (TIP31C, TIP32C).

During project development the robot is powered by a 9 VDC power supply via a flexible umbilical cable.

To test the operation of the H–bridge control circuit, the following sketch is used. The sketch sequentially tests the operation of each motor.

```
//*******************************************************************
//h bridge_test
//*******************************************************************

int lf_fwd = 9, lf_rev = 8, lr_fwd = 7, lr_rev = 6; //robot left signals
int rf_fwd = 5, rf_rev = 4, rr_fwd = 3, rr_rev = 2; //robot right signals

void setup()
{
pinMode(lf_fwd, OUTPUT);    pinMode(lf_rev, OUTPUT); //left fwd motor
pinMode(lr_fwd, OUTPUT);    pinMode(lr_rev, OUTPUT); //left rear motor
pinMode(rf_fwd, OUTPUT);    pinMode(rf_rev, OUTPUT); //right fwd motor
pinMode(rr_fwd, OUTPUT);    pinMode(rr_rev, OUTPUT); //right rear motor
```

---

[5] This platform is available from several different suppliers e.g. Jameco.

[6] Please see www.amazon.com EMOZNY Mecanum wheels with coupling 4 pcs omnidirectional wheels 4 WD smart robot car kit accessories 65 mm robot kit drive wheel for Arduino/Raspberry Pi DIY STEM project.

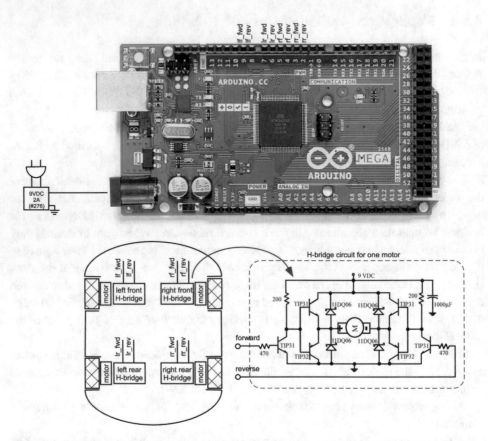

Fig. 4.13 Robot motor control. Mega 2560 R3 illustration used with permission of the Arduino Team (CC BY–NC–SA) (www.arduino.cc)

```
}

void loop()
{
//---Test left front motor-
//forward test
digitalWrite(lf_rev, LOW);      //set reverse signal LOW
analogWrite(lf_fwd, 128);       //analogWrite (0 to 255) set for 50%
delay(2000);                    //delay 2s
//turn off motor
digitalWrite(lf_rev, LOW);      //set reverse signal LOW
digitalWrite(lf_fwd, LOW);      //set forward signal LOW
delay(2000);                    //delay 2s
//reverse test
digitalWrite(lf_fwd, LOW);      //set forward signal LOW
analogWrite(lf_rev, 128);       //analogWrite (0 to 255) set for 50%
delay(2000);                    //delay 2s
```

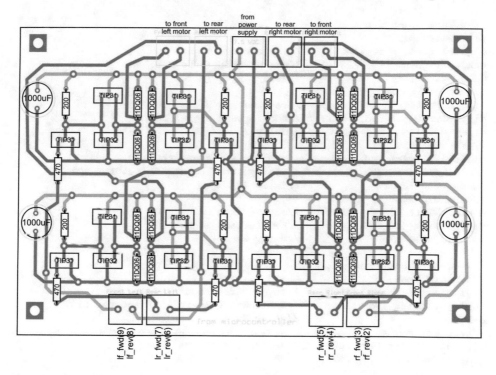

Fig. 4.14 4WD robot PCB

```
//turn off motor
digitalWrite(lf_rev, LOW);     //set reverse signal LOW
digitalWrite(lf_fwd, LOW);     //set forward signal LOW
delay(2000);                   //delay 2s

//---Test left rear motor---
//forward test
digitalWrite(lr_rev, LOW);     //set reverse signal LOW
analogWrite(lr_fwd, 128);      //analogWrite (0 to 255) set for 50%
delay(2000);                   //delay 2s
//turn off motor
digitalWrite(lr_rev, LOW);     //set reverse signal LOW
digitalWrite(lr_fwd, LOW);     //set forward signal LOW
delay(2000);                   //delay 2s
//reverse test
digitalWrite(lr_fwd, LOW);     //set forward signal LOW
analogWrite(lr_rev, 128);      //analogWrite (0 to 255) set for 50%
delay(2000);                   //delay 2s
//turn off motor
digitalWrite(lr_rev, LOW);     //set reverse signal LOW
digitalWrite(lr_fwd, LOW);     //set forward signal LOW
delay(2000);                   //delay 2s
```

```
//---Test right front motor---
//forward test
digitalWrite(rf_rev, LOW);    //set reverse signal LOW
analogWrite(rf_fwd, 128);     //analogWrite (0 to 255) set for 50%
delay(2000);                  //delay 2s
//turn off motor
digitalWrite(rf_rev, LOW);    //set reverse signal LOW
digitalWrite(rf_fwd, LOW);    //set forward signal LOW
delay(2000);                  //delay 2s
//reverse test
digitalWrite(rf_fwd, LOW);    //set forward signal LOW
analogWrite(rf_rev, 128);     //analogWrite (0 to 255) set for 50%
delay(2000);                  //delay 2s
//turn off motor
digitalWrite(rf_rev, LOW);    //set reverse signal LOW
digitalWrite(rf_fwd, LOW);    //set forward signal LOW
delay(2000);                  //delay 2s

//---Test right rear motor---
//forward test
digitalWrite(rr_rev, LOW);    //set reverse signal LOW
analogWrite(rr_fwd, 128);     //analogWrite (0 to 255) set for 50%
delay(2000);                  //delay 2s
//turn off motor
digitalWrite(rr_rev, LOW);    //set reverse signal LOW
digitalWrite(rr_fwd, LOW);    //set forward signal LOW
delay(2000);                  //delay 2s
//reverse test
digitalWrite(rr_fwd, LOW);    //set forward signal LOW
analogWrite(rr_rev, 128);     //analogWrite (0 to 255) set for 50%
delay(2000);                  //delay 2s
//turn off motor
digitalWrite(rr_rev, LOW);    //set reverse signal LOW
digitalWrite(rr_fwd, LOW);    //set forward signal LOW
delay(2000);                  //delay 2s
}

//********************************************************************
```

With the basic testing of the H–bridge network complete, the following options may be pursued:

- Develop functions for the following robot platform actions: turn right, turn left, forward, and reverse. The amount of time the action is rendered should be sent in as a variable to the function.
- Equip the robot platform with Mecanum wheels (Reference Chap. 2). Develop a sketch for the platform to move about a box pattern.

- Develop an optical interface circuit from the Mega 2560 to the four H–bridge circuits.
- Equip the robot platform with optical encoders. See DF Robot Gravity: TT motor encoder kit (#SEN0038).

4.7   Summary

In this chapter we investigated different types of DC motors used onboard robots. The motors may be used for locomotion, sensor positioning and scanning, or actuator positioning. For each type of motor we provided a basic theory of operation, microcontroller interface techniques, and example applications.

4.8   Problems

1. Describe the different types of motors commonly used in robot applications. Provide a suggested application for each type.
2. A stepper motor typically provides 2.5° of rotation per step. How might the resolution of each step be improved (reduced)?
3. Describe the concept of pulse width modulation. How might PWM be used to control the intensity of an LED?
4. In a DC motor what is the relationship between torque and speed?
5. A 9 VDC motor with a stall current of 2 A is to be controlled by an Arduino UNO R3. A 12 VDC, 5 A power supply is available. Design an interface circuit for unidirectional motor movement. Repeat for a bidirectional motor movement.
6. Why are optical isolators required in motor control applications.
7. What is the purpose of the LM324 operational amplifiers in the laser light show circuit.
8. What is the difference between a unipolar and bipolar stepper motor?
9. What controls the speed of rotation of a stepper motor?

References

1. Adafruit homepage, www.adafruit.com
2. *AQV251G PhotoMOS,* www.industrial.panasonic.com, 2021.
3. Arduino homepage, www.arduino.cc
4. Benson, J. (2005). *Next Generation Autonomous Wheelchair Control,* MS thesis, University of Wyoming.
5. Boylestad, R. and L. Nashelsky, (1982). *Electronic Devices and Circuit Theory,* third edition, Prentice–Hall.
6. Clark D. and M. Owings, (2003). *Building Robot Drive Trains,* McGraw Hill.

7. *H11L1M 6–pin DIP Optoisolators Logic Output,* Fairchild Semiconductor Corporation,2003.
8. Fitzgerald, A., C. Kingsley, and S. Umans, (2003). *Electric Machinery*, sixth edition, McGraw Hill.
9. *IRF530 TMOS E–FET Power Field Effect Transistor N–Channel Enhancement–Mode Silicon Gate, (IRF530/D)*, Motorala Inc., 1998.
10. *IRF3205PbF HEXFET Power MOSFET, (PD–94791B)*, International Rectifier: 2010.
11. *IR2104 Half–Bridge Driver (PD60046–S)*, www.irf.com.
12. McMaster–Carr, www.mcmaster.com
13. Sedra, A and K. Smith, (2004). *Microelectronic Circuits*, fifth edition, Oxford University Press.
14. *TIP31C Power Transistors (NPN)*, ST Microelectronics, www.st.com, 2006.
15. *TIP32C Power Transistors (PNP)*, ST Microelectronics, www.st.com, 2006.
16. *TIP120, TIP121, TIP122 (NPN); TIP125, TIP126, TIP127 (PNP) plastic medium–power complementary silicon transistors (TIP120D)*, ON Semiconductor, www.onsemi.com, 2007.

Power Sources

5

Objectives: After reading this chapter, the reader should be able to:

- Describe factors to consider when designing a robot power supply system;
- Specify a power source for an Arduino–based microcontroller board;
- Specify a power source for the payload hosted by a robot; and
- Design and implement a robot power supply system for a specific application.

5.1 Overview

In this brief, but very important chapter, we explore how to design a portable power supply system for a robot. We begin by describing the requirements and parameters to be considered when designing a power supply system. We then investigate different types of battery technologies and related concepts. As discussed in previous chapters, we employ a different power source for the microcontroller and the robot payload to avoid robot motor noise issues. We then design and implement a power source for an Arduino–based microcontroller board followed by a power source for the robot payload. We conclude with a series of examples.[1]

5.2 Project Requirements

An Arduino board is typically used in a wide variety of projects to control external peripheral devices. In this chapter we specifically focus on robot applications. To provide a proper power source for an Arduino–based robot system, the following information must be determined:

---

[1] Some of the information for this chapter is originally from Morgan & Claypool Publishers (M&C) book: "Microcontrollers Fundamentals for Engineers and Scientists."

© The Author(s), under exclusive license to Springer Nature Switzerland AG 2022 189
T. Kerr and S. Barrett, *Arduino IV: DIY Robots*, Synthesis Lectures on Digital Circuits & Systems, https://doi.org/10.1007/978-3-031-11209-6_5

– What is the voltage and current requirements of each device in the system?
– Will the system have any current surge requirements (e.g. motor starting and stall current)?
– Will the system be operated where an AC source is present or will it be a remote system requiring a DC battery supply?
– How long must the system operate before the batteries can be replaced or recharged?
– Is an alternate power source possible (e.g. solar panel)?

Once these questions are answered, a system power supply may be assembled. In the remainder of this chapter, we discuss these different power alternatives. We begin with a review of battery concepts and types.

5.3   Battery Basics

For most applications, robots will carry their own power source in the form of a battery system. In this section we investigate battery basics including ratings, terminology, and types.

5.3.1   Ratings

To select a battery for a robot application, the following requirements must be known: voltage, current, polarity, capacity, and if rechargeable batteries are appropriate for the project (Ada 2021, MIT, Traxxas, Horwitz and Hill, Techlib).

– **voltage:** The unit for voltage is volts. The voltage for a battery is specified for when it is new or fully charged for a rechargeable type battery. Typical battery voltages for common AAA, AA, C, and D non–rechargeable alkaline cells are 1.5 VDC. Rechargeable (secondary) batteries have a slightly lower (1.2–1.25 VDC) terminal voltage.
The voltage of a battery pack may be increased by placing the batteries in series. For example, if a robot requires a 9 VDC battery supply to operate its motors, six D batteries may be placed in series to obtain a 9 VDC battery pack. Plastic battery pack cases are available for battery series stacking. Another common battery type is the 9 VDC rectangular battery with the plus and minus terminals on the same end of the battery. Although this type of battery has the correct voltage it does not have sufficient capacity (described below) to power robot motors.
– **current:** The unit for current is amperes or amps. The current drain of the battery is determined by the load connected to it. For many Arduino based projects the current drain may be specified in milliamps (mA) or Amps (A). As an example, robot drive motors have a rated voltage and also specifications for no load and stall currents. The battery system must be sufficient to supply the current needs of the robot in the most

stressing case (e.g. all four motors running simultaneously for an extended period of time).

The interconnecting wires within a battery pack must be properly gauged to withstand the highest current expected. Wire specifications are provided by the American Wire Gauge (AWG) system. Similarly, a battery pack should be protected against surges and short circuits by appropriately sized fuses.

– **polarity:** In most projects, a positive voltage referenced to ground is required. Some circuits, for example operational amplifier based instrumentation circuits, may require both a positive and negative supply for proper operation. In an upcoming example we demonstrate how to provide a battery based power system with both polarities.

– **capacity:** The battery capacity specification is provided in mAH or AH (amp–hours). It provides an estimate of how long a battery will last under a particular current drain. Common battery capacities are: AAA–1,000 mAH, AA–2,400 mAH, C–6,000 mAH, D–13,000 mAH, and 9 VDC–500 mAH. These values are only approximates. The exact battery capacity is determined by battery technology and manufacturer. Capacity is typically provided with the manufacturer's specification for a battery.

When batteries are placed in series the voltage is the sum of the individual batteries. The capacity of the overall battery pack is not increased. The capacity of the battery pack is the same of an individual battery.

When batteries of the same voltage are placed in parallel, the voltage of the battery back is the same as an individual battery. However, the capacity of the battery pack is the sum of the individual battery capacities within the pack as shown in Fig. 5.1.

– **primary:** Primary batteries are non–rechargeable, single use batteries. We discuss battery types in an upcoming section.

– **secondary:** Secondary batteries are rechargeable and are available in a wide range of voltages and capacities.

– **battery mass and size:** In robot applications, the battery based power system is typically carried onboard the robot. The mass, physical size, and configuration must be accounted for in the robot design.

– **discharge characteristics:** As a battery drains under load, the battery terminal voltage begins to decline. Each battery type has a characteristic slope describing the degradation in battery voltage. Typically, a flat battery discharge profile is desired.

– **voltage regulation:** A battery is typically used with a voltage regulator to maintain the output regulator voltage at a desired level under different current loads. The input voltage to the regulator must typically be several volts higher than the desired output voltage. Voltage regulators are available in a variety of voltage and maximum current ratings.

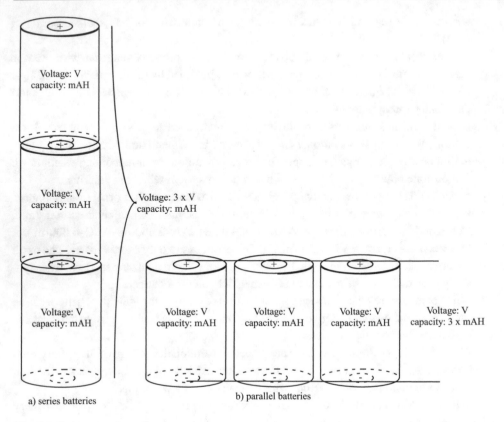

Fig. 5.1 Battery configurations

5.3.2 Types

Provided in Fig. 5.2 is a summary of batteries frequently used in robot applications. Batteries may be separated into primary or secondary types. For each battery type key characteristics are provided.

For larger robot projects, existing power systems may be adapted from surplus powered wheelchairs, rideable children's toys, and golf carts. Typically, these systems use sealed, deep cycle, high capacity (AH rated) lead acid batteries. The batteries are placed in series/parallel combinations to achieve desired system voltage and capacity. Fuses are provided to protect the system.

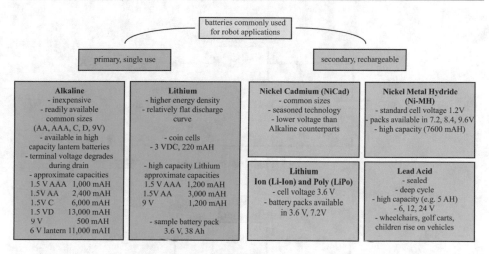

Fig. 5.2 Batteries used in robot applications (Ada 2021, Horowitz and Hill 2015, Techlib, Traxxas)

5.4 Voltage Regulators

As previously mentioned, voltage regulators provide a stabilized, fixed output voltage as the input voltage varies. For example, onboard the Arduino UNO R3 is a voltage regulator providing a fixed output voltage of 5 VDC while the input voltage varies.

A common positive voltage regulator is the 78XX series. The "XX" specifies the regulator output voltage (e.g. 5, 9, 12, etc.). This regulator series has a current rating up to 1.5 A. The input voltage to the regulator typically needs to be several volts higher than the desired output voltage (uA7800).

A common negative voltage regulator is the 79XX series. The "XX" specifies the regulator output voltage (e.g. 5, 9, 12, etc.). This regulator series has a current rating up to 1.5 A. The input voltage to the regulator typically needs to be several volts lower (more negative) than the desired output voltage (LM7900).

Figure 5.3 provides sample circuits to provide a +5 VDC and a ±5 VDC portable battery source.

The LM317 is an adjustable voltage regulator rated at 1.5A. The output voltage can be set for 1.2–37 VDC (LM317). The regulator input voltage must be at least 3 V higher than the desired output voltage as shown in Fig. 5.4. The output voltage value is set choosing the values of R1 and R2 from the following equation (LM317):

$$V_{out} = 1.25V\ (1\ +\ R2/R1)\ +\ I_{adj}R2$$

The second portion of the equation is a small error term.

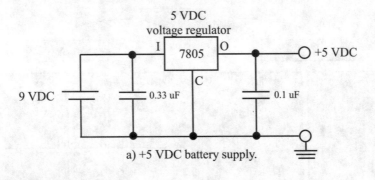

a) +5 VDC battery supply.

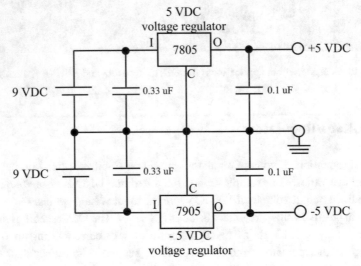

b) +/-5 VDC battery supply.

Fig. 5.3 Battery supply circuits employing a 9 VDC battery with a 5 VDC regulators

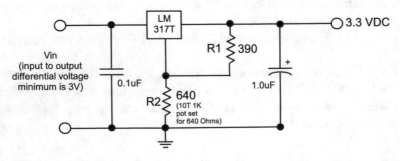

Fig. 5.4 LM317 based 3.3 VDC regulator (LM317)

Example: Design a 3.3 VDC regulator using a LM317 adjustable output positive voltage regulator. In this example we choose R1 to be 390 Ohms. Thus yields a nonstandard resistor value of 640 Ohms for R2. In place of the 640 Ohm resistor, we use a small, ten turn trim potentiometer adjusted such that the regulator output voltage is 3.3 VDC.

5.5   Power Supply System Design

In previous sections we have discussed the power requirements of a microcontroller–based robot system and also the capability of different types of batteries to meet these requirements. To design a battery–based power system, batteries are selected to meet the voltage, capacity, size, and mass requirements of the robot system. In this section we explore battery systems for the microcontroller and the robot payload.

5.5.1   Arduino Power Requirements

Arduino processing boards may be powered from the host computer's USB port during project development. However, it is highly recommended that an external power supply be employed any time a peripheral component is connected. This will allow developing projects beyond the limited current capability of the USB port.

For the UNO and the MEGA platforms, Arduino www.arduino.cc recommends a power supply from 7–12 VDC with a 2.1 mm center positive plug. The Arduino UNO R3 draws approximately 50 mA while the Arduino Mega 2560 R3 draws 75 mA.

Power supplies of this type are readily available from a number of electronic parts supply companies. For example, the Jameco #133891 power supply is a 9 VDC model rated at 300 mA and equipped with a 2.1 mm center positive plug. It is available for under US$10. Power supplies with higher current capacities are readily available. Both the UNO and MEGA have onboard voltage regulators that maintain the incoming power supply voltage to a steady 5 VDC for the onboard processor.

5.5.2   AC Operation via an Umbilical Cable

During project development it is helpful to power the robot project from an AC source. If a source of AC power is readily available, an AC–to–DC converter may be used. These range from a single voltage supply (described above) to a multiple DC voltage power supply with different current specifications for each voltage. When selecting a source it is important to insure it is regulated and fused. As previously discussed, a regulator maintains the source voltage at the same value even under different current loads. A fuse provides protection against a surge current the power supply cannot handle.

a) 9 VDC battery clip   b) 9 VDC battery pack with lid   c) 9 VDC battery pack

Fig. 5.5 Arduino 9 VDC battery power

When the current requirements for each voltage are determined, a power supply may be selected. Choose a power supply with at least double the current specification as required by the maximum demands of the project. Jameco Electronics provides a wide variety of power supplies (www.jameco.com).

5.5.3   Powering the Arduino from Batteries

As previously mentioned, for the UNO and the MEGA platforms, Arduino recommends a power supply from 7–12 VDC with a 2.1 mm center positive plug [www.arduino.cc]. For low power applications a single 9 VDC battery and clip may be used as shown in Fig. 5.5. For higher power applications, a AA battery pack may be used. It is important to note the UNO R3 and Mega R3 Arduino boards have an onboard 5 VDC regulator.

5.5.4   Robot Payload Power Sources

Ideally, the Arduino microcontroller board should be on a separate power supply than the robot payload. Although on separate supplies, they should share a common ground. Because of onboard space limitations, it is not always possible to provide separate supplies.

To design a payload power supply, the voltage and current requirements of each robot system component must be considered. Battery capacity requirements are determined by the mission profile of the robot.

Example: In the Application section of Chap. 4 we equipped a 4WD platform based on the DF Robot 4WD mobile platform chassis with four wheels and bi–directional motor control using four H–bridge channels. Recall, the DF Robot platform is equipped with four powered wheels. The motors are rated at 6 VDC and draw 160 mA each at full speed. The locked rotor current is specified as 2.8 A. In this example we design an AC and DC power supply for the robot as shown in Fig. 5.6.

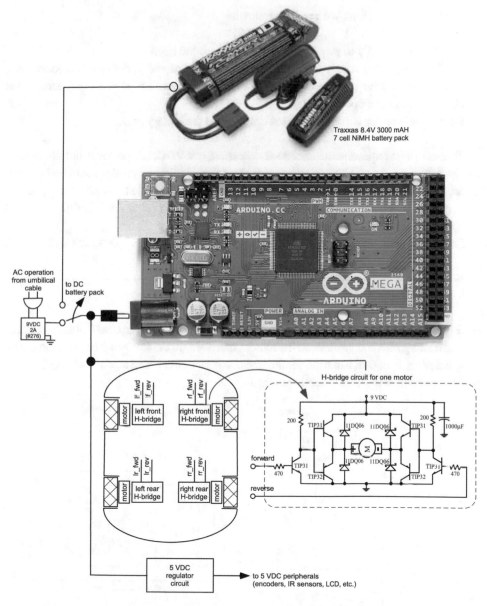

Fig. 5.6 Power system for 4WD mobile platform chassis. Mega 2560 R3 illustration used with permission of the Arduino Team (CC BY–NC–SA) (www.arduino.cc). Battery pack illustration used courtesy of Traxxas (http://m.traxxas.com)

For this robot platform, we need the following:

– A 7–12 VDC supply to provide power for the Arduino board,
– A 5 VDC regulated supply to provide power for the Arduino peripheral components,
– A power supply for the motor (8-9 VDC), capable of drawing 640 mA (or more) when all four motors are operating, and
– Fuse protection for current overloads drawn from the battery pack.

To meet these requirements under AC operation, a 9 VDC, 2A power supply is used as shown in Fig. 5.6. The 9 VDC is delivered to the Arduino microcontroller board and also the H–bridge circuits. The 9 VDC supply is routed to a 5 VDC regulator to provide power for the Arduino peripheral components. Power is provided to the robot via a flexible (stranded wire) umbilical cable.

For battery operation, an 8.4V, 3000 mAH seven cell NiMH battery pack is used [Traxxas].

5.6   Application

In the previous chapter we investigated the importance of providing optical isolation between the controlling microcontroller and the motor controlled. In the following application examples, we provide optical isolation for the linear actuator circuit and an eight channel optical isolator to control the 4WD mobile platform.

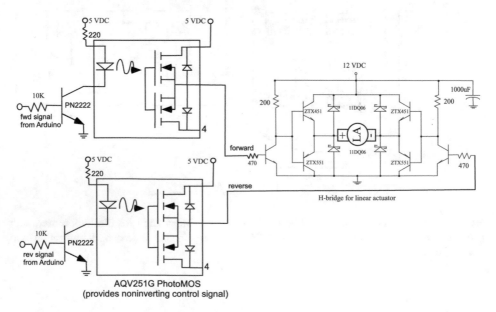

Fig. 5.7 Optical isolation for a small linear actuator

Linear Actuator: In Chap. 4 we equipped a small linear actuator with a discrete H–bridge as shown in Fig. 5.7. The forward and reverse signal from the Arduino controller are routed through an AQV251G PhotoMOS optical isolator and then to the H–bridge.

Eight channel optical isolator: As discussed in Chap. 4, eight different control signals are required to control the four different H–bridges for the 4WD robot platform. As shown in Fig. 5.8, the forward and reverse control signals for a given H–bridge are routed through individual AQV251G PhotoMOS isolators.

Using the information from the linear actuator example, complete the design for all four robot motors.

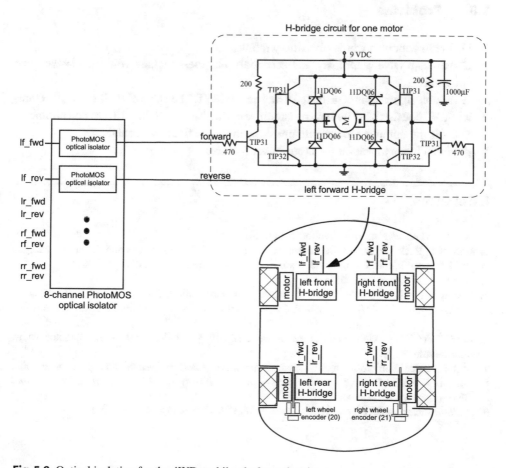

Fig. 5.8 Optical isolation for the 4WD mobile platform chassis

5.7   Summary

In this brief, but very important chapter, we explored how to design a portable power supply system for a robot. We began by describing the requirements and parameters to be considered when designing a power supply system. We then investigated different types of battery technologies and related concepts. As discussed in previous chapters, we employed a different power source for the microcontroller and the robot payload to avoid robot motor noise issues. We then designed and implemented a power source for an Arduino–based microcontroller board followed by a power source for the robot payload.

5.8   Problems

1. Define in your own words key battery ratings.
2. Construct a table of primary and secondary batteries. Include voltage, capacity, and common sizes.
3. A variable regulator is constructed using an LM317T. The value of R1 is set to 390 Ohms while R2 consists of a 100 Ohm resistor in series with a 1,000 Ohm, 10 turn potentiometer. What is the minimum and maximum value available from the regulator?
4. Design a battery pack for the Dagu Rover 5 robot.

References

1. Ada L, *All About Batteries*, www.adafruit.com, 2021
2. Barrett S, Pack D (2006) Microcontrollers Fundamentals for Engineers and Scientists. Morgan and Claypool Publishers. https://doi.org/10.2200/S00025ED1V01Y200605DCS001
3. *Battery Capacity*, www.techlib.com
4. Horowitz P, Hill W (2015) The Art of Electronics, third edition, Cambridge University Press.
5. *LM7095, LM7912, LM7915 LM79XX Series 3–terminal negative regulators*, Texas Instruments, 2013.
6. *LM317, NCV317 1.5A adjustable output, positive voltage regulator*, ON Semiconductor, www.onsemi.com, 2006.
7. MIT Electrical Vehicle Team, *A Guide to Understanding Battery Specifications*, December 2008.
8. *Traxxas Guide to Batteries and Chargers–Learn the Basics About Battery Types, Ratings, and Charging*, www.traxxas.com
9. *uA7800 series positive–voltage regulators*, SLVS056J, Texas Instruments, 2003.

Applications

Objectives: After reading this chapter, the reader should be able to:

- Design, 3D print, implement, and test a 4WD maze navigating robot.

6.1 Overview

The purpose of this chapter is to put into practice lessons learned throughout the book. Due to space constraints we limit our exploration to only one project. We considered a number of motivating alternatives including a robot fish, a robot snake, a bionic artificial hand, and a 4WD maze navigating robot. We chose the 4WD maze navigating robot to demonstrate how a number of sensors and actuators may be used in a single, integrated system.

6.2 Mountain Maze Navigating Robot

In this project we extend the maze navigating robot project introduced early in the book to a three–dimensional mountain pass robot maze. We use a robot equipped with four motorized wheels. Each of the wheels is equipped with an H–bridge to allow bidirectional motor control. The robot is also equipped with additional instrumentation.

© The Author(s), under exclusive license to Springer Nature Switzerland AG 2022
T. Kerr and S. Barrett, *Arduino IV: DIY Robots*, Synthesis Lectures on Digital Circuits & Systems, https://doi.org/10.1007/978-3-031-11209-6_6

6.2.1   Description

For this project the DF Robot 4WD mobile platform kit is used (DFROBOT ROB0003, Jameco #2124285).[1] The robot kit is equipped with four powered wheels. We equip the robot with three Sharp GP2Y0A41SK0F IR sensors to sense maze walls. The robot is placed in a three dimensional maze with reflective walls modeled after a mountain pass. The goal of the project is for the robot to detect wall placement and navigate through the maze. The robot will not be provided any information about the maze. The control algorithm for the robot is hosted on the Arduino Mega 2560 Rev 3.

6.2.2   Requirements

The requirements for this project are simple: the robot must autonomously navigate through the maze without touching maze walls.

6.2.3   Circuit Diagram

The circuit diagram for the robot is provided in Fig. 6.1. The three IR sensors (left, center, and right) are mounted on the leading edge of the robot to detect maze walls. The output from the sensors are fed to three Mega 2560 ADC channels (A0, A1, and A2). The robot motors are controlled by eight different PWM channels (lf_fwd(9), lf_rev(8), lr_fwd(7) lr_rev(6), rf_fwd(5), rf_rev(4), rr_fwd(3), and rr_rev(2)) via four independent H–bridges as discussed in previous chapters.

The robot is powered by an 8.4 VDC NiMH battery pack rated at 3000 mAH as discussed in the previous chapter. The battery voltage is fed to the Arduino Mega 2560 Rev 3, the four H–bridges and to a 5 VDC voltage regulator. Alternatively, the robot may be powered by a 9 VDC power supply rated at several amps. In this case the power is delivered to the robot by a flexible umbilical cable.

6.2.4   Structure Chart

The structure chart for the robot project is provided in Fig. 6.2.

---

[1] Variations of this project have been used in several Morgan and Claypool publications. We adapt the project to demonstrate the powerful features of the Arduino Mega 2560 Rev 3.

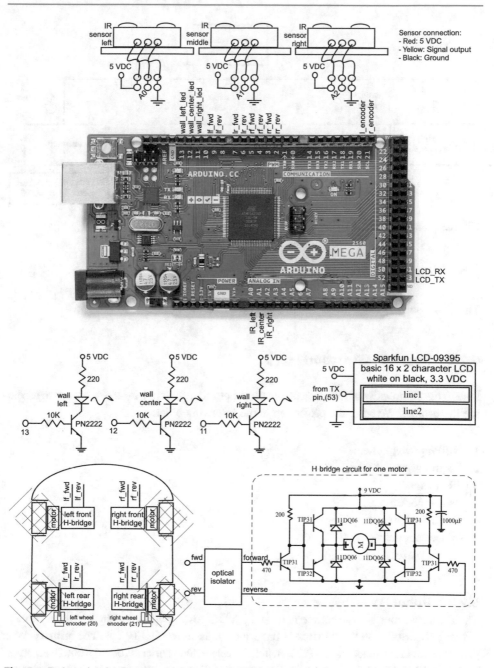

Fig. 6.1 Robot circuit diagram. Arduino Mega 2560 R3 illustration used with permission of the Arduino Team (CC BY–NC–SA) (www.arduino.cc)

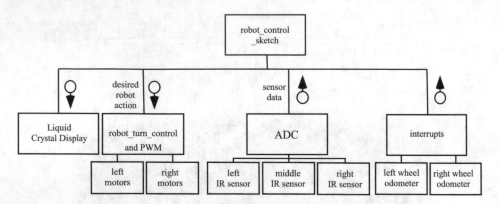

Fig. 6.2 Robot structure diagram

6.2.5   UML Activity Diagrams

The UML activity diagram for the robot is provided in Fig. 6.3.

6.2.6   Robot Construction

Due to the complexity of this project, we use a bottom–up, step–by–step approach to assemble the project. We bring the project up in the following order:

1. Robot power system
2. Liquid crystal display
3. IR sensors
4. Motor control
5. Control algorithm
6. Wheel odometry
7. Robot maze

6.2.6.1 Robot Power System

The primary robot voltage source is from a 9 VDC power supply via an umbilical cable or via an onboard 8.4 VDC, 3000 mAH battery pack as shown in Fig. 6.4. The primary robot power source is selected via a SPDT switch. The selected source is routed through an on/off switch, a 12 VDC 2A fuse, and a 5 VDC regulator. The selected source is also routed to the robot chassis to provide motor power.

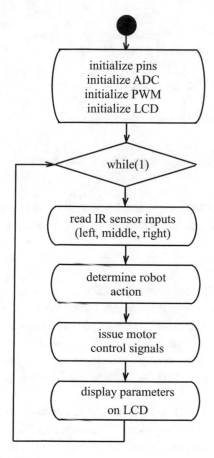

Fig. 6.3 Robot UML activity diagram

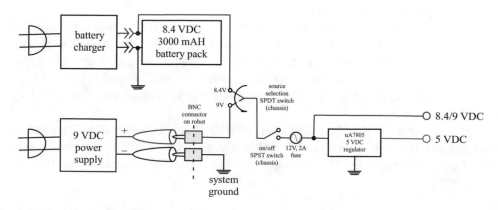

Fig. 6.4 Robot power system

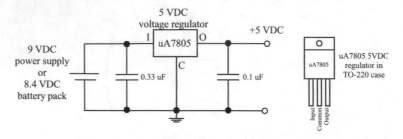

Fig. 6.5 Battery supply circuits employing a 8.4 or 9 VDC source with a 5 VDC regulator

A 5 VDC regulator, uA7805, rated at 1.5 A is used to provide a regulated 5 VDC supply for the peripheral components (e.g. LCD, IR sensors, LEDs, etc.). The regulator circuit is shown in Fig. 6.5.

6.2.6.2 Serial Liquid Crystal Display (LCD)

An LCD[2] is an output device to display text information. LCDs come in a wide variety of configurations including multi–character, multi–line format. A 16×2 LCD format is common. That is, it has the capability of displaying two lines of 16 characters each. Each display character and line has a specific associated address. The characters are sent to the LCD via American Standard Code for Information Interchange (ASCII) format a single character at a time.

For a parallel configured LCD, an eight bit data path and two lines are required between the microcontroller and the LCD. Many parallel configured LCDs may also be configured for a four–bit data path thus saving several precious microcontroller pins. A small microcontroller mounted to the back panel of the LCD translates the ASCII data characters and control signals to properly display the characters.

To conserve precious, limited microcontroller input/output pins, a serial configured LCD is used in this project. A serial LCD reduces the number of required microcontroller pins for interface, from ten down to one, as shown in Fig. 6.6. Display data and control information is sent to the LCD via an asynchronous UART serial communication link (8 data bits, 1 stop bit, no parity, 9600 Baud). A serial configured LCD costs slightly more than a similarly configured parallel LCD.

For this project, a SparkFun LCD–09395, 5.0 VDC, serial, 16 by 2 character, black on white LCD display is connected to the Arduino Mega 2560 R3. Communication between the Mega 2560 and the LCD is accomplished by a single 9600 bits per second (BAUD) connection.

---

[2] This section was adapted from "Arduino I: Getting Started, S. Barrett.

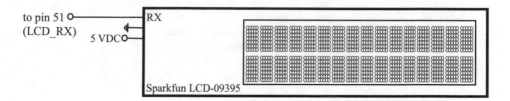

Line	Character Position (n)
1	0-15
2	64-79

Note: character position is specifed
as $0X80 + n$

Command Code	Command
0x01	Clear Display
0x14	Cursor one space right
0x10	Cursor one space left
0x80 ⏐ n	Cursor to position

Note: precede command with 0xFE (254_{10})

Fig. 6.6 LCD serial display

Rather than use the onboard Universal Asynchronous Receiver Transmitter (UART), the Arduino Software Serial Library is used. The library provides functions to mimic UART activities on a digital pin. Details on the Library are provided at the Arduino website (www. arduino.cc).

```
//***************************************************************
//LCD_robot
//Example uses the Arduino Software Serial Library with the
//SparkFun LCD-09395.
// - provides software-based serial port
//***************************************************************

#include <SoftwareSerial.h>

//Specify Arduino pins for Serial connection:
//  SoftwareSerial LCD(RX_pin, TX_pin);
SoftwareSerial LCD(51, 53);

void setup()
{
LCD.begin(9600);                    //Baud rate: 9600 Baud
delay(500);                         //Delay for display
}

void loop()
{
```

```
//Cursor to line one, character one
LCD.write(254);                         //Command prefix
LCD.write(128);                         //Command

//clear display
LCD.write("                    ");
LCD.write("                    ");

//Cursor to line one, character one
LCD.write(254);                         //Command prefix
LCD.write(128);                         //Command

LCD.write("SerLCD Test");

//Cursor to line two, character one
LCD.write(254);                         //Command prefix
LCD.write(192);                         //Command

LCD.write("LCD-09395");

while(1);                               //pause here
}

//********************************************************************
```

6.2.6.3 IR Sensors

In this step we equip the robot with three Sharp GP2Y0A41SK0F (GP2) IR sensors to sense maze walls at the robot's left, center, and right. The sensor outputs are routed to the Mega 2560's analog inputs (A0, A1, and A2). Maze walls are detected when the sensor voltage reaches a desired, preset level. When a wall is detected, a corresponding LED is illuminated.

```
//********************************************************************
//robot_sensors
//Three IR sensors (left, middle, and right) are mounted on the leading
//edge of the robot to detect maze walls.  The sensors' outputs are
//fed to three separate ADC channels on A0, A1, and A2.
//
//The robot is equipped with a SparkFun LCD-09395
//  - Uses Arduino SoftwareSerial library
//  - 51: LCD_RX
//  - 53: LCD_TX
//LEDs to indicate wall detection are at 13, 12, 11.
//
//This example code is in the public domain.
//********************************************************************

#include <SoftwareSerial.h>

//Specify Arduino pins for Serial connection:
//  SoftwareSerial LCD(RX_pin, TX_pin);
```

```
SoftwareSerial LCD(51, 53);


                                      //analog input pins
#define left_IR_sensor     A0         //analog pin - left IR sensor
#define center_IR_sensor   A1         //analog pin - center IR sensor
#define right_IR_sensor    A2         //analog pin - right IR sensor


                                      //digital output pins
                                      //LED indicators - wall detectors
#define wall_left          13         //digital pin - wall_left
#define wall_center        12         //digital pin - wall_center
#define wall_right         11         //digital pin - wall_right
                                      //sensor value
int left_IR_sensor_value;             //variable for left IR sensor
int center_IR_sensor_value;           //variable for center IR sensor
int right_IR_sensor_value;            //variable for right IR sensor

int troubleshoot;                     //asserts troubleshoot statements

void setup()
{
LCD.begin(9600);                      //Baud rate: 9600 Baud
delay(500);                           //Delay for display
                                      //LED indicators - wall detectors
pinMode(wall_left,   OUTPUT);         //configure pin for digital output
pinMode(wall_center, OUTPUT);         //configure pin for digital output
pinMode(wall_right,  OUTPUT);         //configure pin for digital output
}

void loop()
{
//read analog output from IR sensors
left_IR_sensor_value   = analogRead(left_IR_sensor);
center_IR_sensor_value = analogRead(center_IR_sensor);
right_IR_sensor_value  = analogRead(right_IR_sensor);

//Clear LCD
//Cursor to line one, character one
LCD.write(254);                       //Command prefix
LCD.write(128);                       //Command

//clear display
LCD.write("               ");
LCD.write("               ");

//Cursor to line one, character one
LCD.write(254);                       //Command prefix
LCD.write(128);                       //Command
LCD.write("Left  Ctr  Right");
delay(50);

LCD.write(254);                       //Command to LCD
delay(5);
```

```
LCD.write(192);                    //Cursor to line 2, position 1
delay(5);
LCD.print(left_IR_sensor_value);
delay(5);
LCD.write(254);                    //Command to LCD
delay(5);
LCD.write(198);                    //Cursor to line 2, position 8
delay(5);
LCD.print(center_IR_sensor_value);
delay(5);
LCD.write(254);                    //Command to LCD
delay(5);
LCD.write(203);                    //Cursor to line 2, position 13
delay(5);
LCD.print(right_IR_sensor_value);
delay(5);
delay(500);

//robot action table row 0 - robot forward
if((left_IR_sensor_value < 300)&&(center_IR_sensor_value < 300)&&
  (right_IR_sensor_value < 300))
  {
                                        //wall detection LEDs
  digitalWrite(wall_left,   LOW);       //turn LED off
  digitalWrite(wall_center, LOW);       //turn LED off
  digitalWrite(wall_right,  LOW);       //turn LED off

  //desired robot action
  }

//robot action table row 1 - robot forward
else if((left_IR_sensor_value < 300)&&(center_IR_sensor_value < 300)&&
      (right_IR_sensor_value > 300))
  {
                                        //wall detection LEDs
  digitalWrite(wall_left,   LOW);       //turn LED off
  digitalWrite(wall_center, LOW);       //turn LED off
  digitalWrite(wall_right,  HIGH);      //turn LED on

  //desired robot action
  }

//robot action table row 2 - robot right
else if((left_IR_sensor_value < 300)&&(center_IR_sensor_value > 300)&&
      (right_IR_sensor_value < 300))
  {
                                        //wall detection LEDs
  digitalWrite(wall_left,   LOW);       //turn LED off
  digitalWrite(wall_center, HIGH);      //turn LED on
  digitalWrite(wall_right,  LOW);       //turn LED off

  //desired robot action
  }
```

```
//robot action table row 3 - robot left
else if((left_IR_sensor_value < 300)&&(center_IR_sensor_value > 300)&&
        (right_IR_sensor_value > 300))
   {
                                          //wall detection LEDs
   digitalWrite(wall_left,   LOW);        //turn LED off
   digitalWrite(wall_center, HIGH);       //turn LED on
   digitalWrite(wall_right,  HIGH);       //turn LED on

   //desired robot action
   }

//robot action table row 4 - robot forward
else if((left_IR_sensor_value > 300)&&(center_IR_sensor_value < 300)&&
        (right_IR_sensor_value < 300))
   {
                                          //wall detection LEDs
   digitalWrite(wall_left,   HIGH);       //turn LED on
   digitalWrite(wall_center, LOW);        //turn LED off
   digitalWrite(wall_right,  LOW);        //turn LED off

   //desired robot action
   }

//robot action table row 5 - robot forward
else if((left_IR_sensor_value > 300)&&(center_IR_sensor_value < 300)&&
        (right_IR_sensor_value > 300))
   {
                                          //wall detection LEDs
   digitalWrite(wall_left,   HIGH);       //turn LED on
   digitalWrite(wall_center, LOW);        //turn LED off
   digitalWrite(wall_right,  HIGH);       //turn LED on

   //desired robot action
   }

//robot action table row 6 - robot right
else if((left_IR_sensor_value > 300)&&(center_IR_sensor_value > 300)&&
        (right_IR_sensor_value < 300))
   {
                                          //wall detection LEDs
   digitalWrite(wall_left,   HIGH);       //turn LED on
   digitalWrite(wall_center, HIGH);       //turn LED on
   digitalWrite(wall_right,  LOW);        //turn LED off

   //desired robot action
   }

//robot action table row 7 - robot reverse
else if((left_IR_sensor_value > 300)&&(center_IR_sensor_value > 300)&&
        (right_IR_sensor_value > 300))
   {
```

```
                                               //wall detection LEDs
  digitalWrite(wall_left,   HIGH);             //turn LED on
  digitalWrite(wall_center, HIGH);             //turn LED on
  digitalWrite(wall_right,  HIGH);             //turn LED on

  //desired robot action
  }
}


//*********************************************************************
```

6.2.6.4 Motor Control

In the Application section of Chap. 4, we equipped the DF Robot 4WD platform with four independently controlled, bidirectional motors. We use this example to provide motor control for this project. Selected portions are repeated here for completeness.

The circuit diagram for the robot equipped with four H–bridges to independently control the direction and speed of each motor is provided in Fig. 4.13. The four robot motors are driven by eight separate PWM channels from Mega 2560 pins via four independent H–

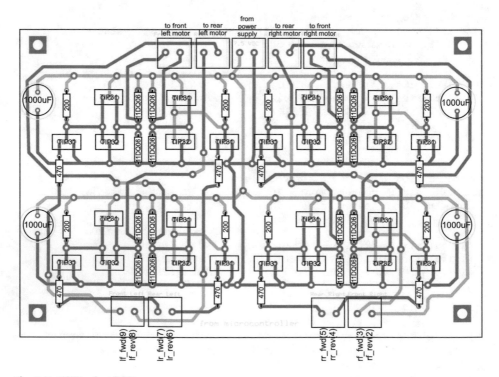

Fig. 6.7 4WD robot PCB

bridges. To operate a given motor in the forward direction, a PWM signal of desired duty cycle to provide desired motor speed is provided to the "fwd" signal of the H–bridge (e.g. lf_fwd). At the same time, a logic low is provided to the "rev" (e.g. lf_rev) input of the same H–bridge. Alternatively, to operate a motor in the reverse direction, a PWM signal of desired duty cycle to provide desired motor speed is provided to the "rev" signal of the H–bridge (e.g. lf_rev). Simultaneously, a logic low is provided to the "fwd" (e.g. lf_fwd) input of the same H–bridge (Fig. 6.7).

To test the operation of the H–bridge control circuit, the following script is used:

```
//******************************************************************
//h_bridge_test
//******************************************************************

int lf_fwd = 9, lf_rev = 8, lr_fwd = 7, lr_rev = 6; //robot left signals
int rf_fwd = 5, rf_rev = 4, rr_fwd = 3, rr_rev = 2; //robot right signals

void setup()
{
pinMode(lf_fwd, OUTPUT);      pinMode(lf_rev, OUTPUT); //left fwd motor
pinMode(lr_fwd, OUTPUT);      pinMode(lr_rev, OUTPUT); //left rear motor
pinMode(rf_fwd, OUTPUT);      pinMode(rf_rev, OUTPUT); //right fwd motor
pinMode(rr_fwd, OUTPUT);      pinMode(rr_rev, OUTPUT); //right rear motor
}

void loop()
{
//---Test left front motor-
//forward test
digitalWrite(lf_rev, LOW);    //set reverse signal LOW
analogWrite(lf_fwd, 128);     //analogWrite (0 to 255) set for 50%
delay(2000);                  //delay 2s
//turn off motor
digitalWrite(lf_rev, LOW);    //set reverse signal LOW
digitalWrite(lf_fwd, LOW);    //set forward signal LOW
delay(2000);                  //delay 2s
//reverse test
digitalWrite(lf_fwd, LOW);    //set forward signal LOW
analogWrite(lf_rev, 128);     //analogWrite (0 to 255) set for 50%
delay(2000);                  //delay 2s
//turn off motor
digitalWrite(lf_rev, LOW);    //set reverse signal LOW
digitalWrite(lf_fwd, LOW);    //set forward signal LOW
delay(2000);                  //delay 2s

//---Test left rear motor---
//forward test
digitalWrite(lr_rev, LOW);    //set reverse signal LOW
analogWrite(lr_fwd, 128);     //analogWrite (0 to 255) set for 50%
delay(2000);                  //delay 2s
//turn off motor
digitalWrite(lr_rev, LOW);    //set reverse signal LOW
```

```
digitalWrite(lr_fwd, LOW);      //set forward signal LOW
delay(2000);                    //delay 2s
//reverse test
digitalWrite(lr_fwd, LOW);      //set forward signal LOW
analogWrite(lr_rev, 128);       //analogWrite (0 to 255) set for 50%
delay(2000);                    //delay 2s
//turn off motor
digitalWrite(lr_rev, LOW);      //set reverse signal LOW
digitalWrite(lr_fwd, LOW);      //set forward signal LOW
delay(2000);                    //delay 2s

//---Test right front motor---
//forward test
digitalWrite(rf_rev, LOW);      //set reverse signal LOW
analogWrite(rf_fwd, 128);       //analogWrite (0 to 255) set for 50%
delay(2000);                    //delay 2s
//turn off motor
digitalWrite(rf_rev, LOW);      //set reverse signal LOW
digitalWrite(rf_fwd, LOW);      //set forward signal LOW
delay(2000);                    //delay 2s
//reverse test
digitalWrite(rf_fwd, LOW);      //set forward signal LOW
analogWrite(rf_rev, 128);       //analogWrite (0 to 255) set for 50%
delay(2000);                    //delay 2s
//turn off motor
digitalWrite(rf_rev, LOW);      //set reverse signal LOW
digitalWrite(rf_fwd, LOW);      //set forward signal LOW
delay(2000);                    //delay 2s

//---Test right rear motor---
//forward test
digitalWrite(rr_rev, LOW);      //set reverse signal LOW
analogWrite(rr_fwd, 128);       //analogWrite (0 to 255) set for 50%
delay(2000);                    //delay 2s
//turn off motor
digitalWrite(rr_rev, LOW);      //set reverse signal LOW
digitalWrite(rr_fwd, LOW);      //set forward signal LOW
delay(2000);                    //delay 2s
//reverse test
digitalWrite(rr_fwd, LOW);      //set forward signal LOW
analogWrite(rr_rev, 128);       //analogWrite (0 to 255) set for 50%
delay(2000);                    //delay 2s
//turn off motor
digitalWrite(rr_rev, LOW);      //set reverse signal LOW
digitalWrite(rr_fwd, LOW);      //set forward signal LOW
delay(2000);                    //delay 2s
}

//*********************************************************************
```

6.2.6.5 Control Algorithm

To control robot action, the information gathered from the three IR sensors must be translated to desired robot motor action. Figure 6.8 summarizes the desired robot action. The following Arduino script implements these actions.

```
//**********************************************************************
//robot_control
//Three IR sensors (left, middle, and right) are mounted on the leading
//edge of the robot to detect maze walls.  The sensors' outputs are
//fed to three separate ADC channels on A0, A1, and A2.
//
//The robot is equipped with a SparkFun LCD-09395
//   - Uses Arduino SoftwareSerial library
//   - 51: LCD_RX
//   - 53: LCD_TX
//LEDs to indicate wall detection are at 13, 12, 11.
//
//For robot movements, the wheels/motors are paired on the left and
//the right.
//
//This example code is in the public domain.
//**********************************************************************

#include <SoftwareSerial.h>

//Specify Arduino pins for Serial connection:
//   SoftwareSerial LCD(RX_pin, TX_pin);
SoftwareSerial LCD(51, 53);


                                     //analog input pins
#define left_IR_sensor      A0       //analog pin - left IR sensor
#define center_IR_sensor    A1       //analog pin - center IR sensor
#define right_IR_sensor     A2       //analog pin - right IR sensor


                                     //digital output pins
                                     //LED indicators - wall detectors
#define wall_left           13       //digital pin - wall_left
#define wall_center         12       //digital pin - wall_center
#define wall_right          11       //digital pin - wall_right
                                     //sensor value
int left_IR_sensor_value;            //variable for left IR sensor
int center_IR_sensor_value;          //variable for center IR sensor
int right_IR_sensor_value;           //variable for right IR sensor
int lf_fwd = 9, lf_rev = 8, lr_fwd = 7, lr_rev = 6;   //robot left controls
int rf_fwd = 5, rf_rev = 4, rr_fwd = 3, rr_rev = 2;   //robot right controls


void setup()
{
LCD.begin(9600);                     //Baud rate: 9600 Baud
delay(500);                          //Delay for display
                                         //LED indicators - wall detectors
```

```
pinMode(wall_left,   OUTPUT);          //configure pin for digital output
pinMode(wall_center, OUTPUT);          //configure pin for digital output
pinMode(wall_right,  OUTPUT);          //configure pin for digital output

pinMode(lf_fwd, OUTPUT);     pinMode(lf_rev, OUTPUT);//left forward motor
pinMode(lr_fwd, OUTPUT);     pinMode(lr_rev, OUTPUT);//left rear motor
pinMode(rf_fwd, OUTPUT);     pinMode(rf_rev, OUTPUT);//right forward motor
pinMode(rr_fwd, OUTPUT);     pinMode(rr_rev, OUTPUT);//right rear motor
}

void loop()
{
//read analog output from IR sensors
left_IR_sensor_value   = analogRead(left_IR_sensor);
center_IR_sensor_value = analogRead(center_IR_sensor);
right_IR_sensor_value  = analogRead(right_IR_sensor);

//Clear LCD
//Cursor to line one, character one
LCD.write(254);                        //Command prefix
LCD.write(128);                        //Command

//clear display
LCD.write("                 ");
LCD.write("                 ");

//Cursor to line one, character one
LCD.write(254);                        //Command prefix
LCD.write(128);                        //Command
LCD.write("Left  Ctr  Right");
delay(50);

LCD.write(254);                 //Command to LCD
delay(5);
LCD.write(192);                 //Cursor to line 2, position 1
delay(5);
LCD.print(left_IR_sensor_value);
delay(5);
LCD.write(254);                 //Command to LCD
delay(5);
LCD.write(198);                 //Cursor to line 2, position 8
delay(5);
LCD.print(center_IR_sensor_value);
delay(5);
LCD.write(254);                 //Command to LCD
delay(5);
LCD.write(203);                 //Cursor to line 2, position 13
delay(5);
LCD.print(right_IR_sensor_value);
delay(5);
delay(500);

//robot action table row 0 - robot forward
```

```
if((left_IR_sensor_value < 300)&&(center_IR_sensor_value < 300)&&
   (right_IR_sensor_value < 300))
   {
                                          //wall detection LEDs
   digitalWrite(wall_left,   LOW);        //turn LED off
   digitalWrite(wall_center, LOW);        //turn LED off
   digitalWrite(wall_right,  LOW);        //turn LED off

   //desired robot action - row 0 - robot forward
   robot_forward(2000);
   robot_stop();
   }

//robot action table row 1 - robot forward
else if((left_IR_sensor_value < 300)&&(center_IR_sensor_value < 300)&&
        (right_IR_sensor_value > 300))
   {
                                          //wall detection LEDs
   digitalWrite(wall_left,   LOW);        //turn LED off
   digitalWrite(wall_center, LOW);        //turn LED off
   digitalWrite(wall_right,  HIGH);       //turn LED on

   //desired robot action  - row 1 - robot forward
   robot_forward(2000);
   robot_stop();
   }

//robot action table row 2 - robot right
else if((left_IR_sensor_value < 300)&&(center_IR_sensor_value > 300)&&
        (right_IR_sensor_value < 300))
   {
                                          //wall detection LEDs
   digitalWrite(wall_left,   LOW);        //turn LED off
   digitalWrite(wall_center, HIGH);       //turn LED on
   digitalWrite(wall_right,  LOW);        //turn LED off

   //desired robot action - row 2 - robot right
   robot_right(2000);
   robot_stop();
   }

//robot action table row 3 - robot left
else if((left_IR_sensor_value < 300)&&(center_IR_sensor_value > 300)&&
        (right_IR_sensor_value > 300))
   {
                                          //wall detection LEDs
   digitalWrite(wall_left,   LOW);        //turn LED off
   digitalWrite(wall_center, HIGH);       //turn LED on
   digitalWrite(wall_right,  HIGH);       //turn LED on

   //desired robot action - row 3 - robot left
   robot_left(2000);
   robot_stop();
```

```
  }

//robot action table row 4 - robot forward
else if((left_IR_sensor_value > 300)&&(center_IR_sensor_value < 300)&&
        (right_IR_sensor_value < 300))
  {
                                          //wall detection LEDs
  digitalWrite(wall_left,   HIGH);        //turn LED on
  digitalWrite(wall_center, LOW);         //turn LED off
  digitalWrite(wall_right,  LOW);         //turn LED off

  //desired robot action - row 4 - robot forward
  robot_forward(2000);
  robot_stop();
  }

//robot action table row 5 - robot forward
else if((left_IR_sensor_value > 300)&&(center_IR_sensor_value < 300)&&
        (right_IR_sensor_value > 300))
  {
                                          //wall detection LEDs
  digitalWrite(wall_left,   HIGH);        //turn LED on
  digitalWrite(wall_center, LOW);         //turn LED off
  digitalWrite(wall_right,  HIGH);        //turn LED on

  //desired robot action - row 4 - robot forward
  robot_forward(2000);
  robot_stop();
  }

//robot action table row 6 - robot right
else if((left_IR_sensor_value > 300)&&(center_IR_sensor_value > 300)&&
        (right_IR_sensor_value < 300))
  {
                                          //wall detection LEDs
  digitalWrite(wall_left,   HIGH);        //turn LED on
  digitalWrite(wall_center, HIGH);        //turn LED on
  digitalWrite(wall_right,  LOW);         //turn LED off

  //desired robot action - row 6 - robot right
  robot_right(2000);
  robot_stop();
  }

//robot action table row 7 - robot reverse
else if((left_IR_sensor_value > 300)&&(center_IR_sensor_value > 300)&&
        (right_IR_sensor_value > 300))
  {
                                          //wall detection LEDs
  digitalWrite(wall_left,   HIGH);        //turn LED on
  digitalWrite(wall_center, HIGH);        //turn LED on
  digitalWrite(wall_right,  HIGH);        //turn LED on
```

```
    //desired robot action - row 7 - robot reverse
    robot_reverse(2000);
    robot_stop();
    }
}

//*************************************************************************
//robot forward - all wheels forward for specified delay
//*************************************************************************

void robot_forward(unsigned int motor_time)
{
//left front motor (lf)
digitalWrite(lf_rev, LOW);     //set reverse signal LOW
analogWrite(lf_fwd, 128);      //analogWrite values from 0 to 255     50%

//left rear motor (lr)
digitalWrite(lr_rev, LOW);     //set reverse signal LOW
analogWrite(lr_fwd, 128);      //analogWrite values from 0 to 255     50%

//right front motor (rf)
digitalWrite(rf_rev, LOW);     //set reverse signal LOW
analogWrite(rf_fwd, 128);      //analogWrite values from 0 to 255     50%

//right rear motor (rr)
digitalWrite(rr_rev, LOW);     //set reverse signal LOW
analogWrite(rr_fwd, 128);      //analogWrite values from 0 to 255     50%

delay(motor_time);
}

//*************************************************************************
//robot stop - all wheels stop
//*************************************************************************

void robot_stop(void)
{
//left front motor (lf)
digitalWrite(lf_rev, LOW);     //set reverse signal LOW
analogWrite(lf_fwd, LOW);      //set forward signal LOW

//left rear motor (lr)
digitalWrite(lr_rev, LOW);     //set reverse signal LOW
analogWrite(lr_fwd, LOW);      //set forward signal LOW

//right front motor (rf)
digitalWrite(rf_rev, LOW);     //set reverse signal LOW
analogWrite(rf_fwd, LOW);      //set forward signal LOW

//right rear motor (rr)
digitalWrite(rr_rev, LOW);     //set reverse signal LOW
analogWrite(rr_fwd, LOW);      //set forward signal LOW
}
```

```
//***********************************************************************
//robot reverse - left wheels forward, right wheels reverse for
//                specified delay
//***********************************************************************

void robot_reverse(unsigned int motor_time)
{
//left front motor (lf)
digitalWrite(lf_rev, LOW);    //set reverse signal LOW
analogWrite(lf_fwd, 128);     //analogWrite values from 0 to 255    50%

//left rear motor (lr)
digitalWrite(lr_rev, LOW);    //set reverse signal LOW
analogWrite(lr_fwd, 128);     //analogWrite values from 0 to 255    50%

//right front motor (rf)
digitalWrite(rf_rev, 128);    //analogWrite values from 0 to 255    50%
analogWrite(rf_fwd, LOW);     //set forward signal LOW

//right rear motor (rr)
digitalWrite(rr_rev, 128);    //analogWrite values from 0 to 255    50%
analogWrite(rr_fwd, LOW);     //set forward signal LOW

delay(motor_time);
}

//***********************************************************************
//robot_left - left wheels reverse, right wheels forward for
//                specified delay
//***********************************************************************

void robot_left(unsigned int motor_time)
{
//left front motor (lf)
digitalWrite(lf_rev, 128);    //analogWrite values from 0 to 255    50%
analogWrite(lf_fwd, LOW);     //set forward signal LOW

//left rear motor (lr)
digitalWrite(lr_rev, 128);    //analogWrite values from 0 to 255    50%
analogWrite(lr_fwd, LOW);     //set forward signal LOW

//right front motor (rf)
digitalWrite(rf_rev, LOW);    //set reverse signal LOW
analogWrite(rf_fwd, 128);     //analogWrite values from 0 to 255    50%

//right rear motor (rr)
digitalWrite(rr_rev, LOW);    //set reverse signal LOW
analogWrite(rr_fwd, 128);     //analogWrite values from 0 to 255    50%

delay(motor_time);
}
```

```
//**********************************************************************
//robot right - left wheels forward, right wheels reverse for
//                  specified delay
//**********************************************************************

void robot_right(unsigned int motor_time)
{
//left front motor (lf)
digitalWrite(lf_rev, LOW);      //set reverse signal LOW
analogWrite(lf_fwd, 128);       //analogWrite values from 0 to 255    50%

//left rear motor (lr)
digitalWrite(lr_rev, LOW);      //set reverse signal LOW
analogWrite(lr_fwd, 128);       //analogWrite values from 0 to 255    50%

//right front motor (rf)
digitalWrite(rf_rev, 128);      //analogWrite values from 0 to 255    50%
analogWrite(rf_fwd, LOW);       //set forward signal LOW

//right rear motor (rr)
digitalWrite(rr_rev, 128);      //analogWrite values from 0 to 255    50%
analogWrite(rr_fwd, LOW);       //set forward signal LOW

delay(motor_time);
}

//**********************************************************************
```

6.2.6.6 Wheel Odometry

In Chap. 3 we discussed wheel odometry in some detail. For this project we equip the back left and back right motors with a DF Robot Gravity: TT motor encoder kit as shown in Fig. 6.9. The encoders operate on 5 VDC, drawing 20 mA, and provide 20 pulses per wheel revolution. We leave integrating the encoders as a project extension assignment.

6.2.7   Robot Chassis–Earth Roamer

On the way to work one morning, I (sfb) pulled into a local gas station to fill up my gas tank. Next to my car was an Earth Roamer Recreational Vehicle (https://earthroamer.com). I had never seen such a magnificent vehicle. When Tyler and I met to discuss the chassis for this project, the Earth Roamer came to mind. In this section, you see Tyler's genius at work as he conceptualized, designed, and 3D printed the Earth Roamer inspired chassis for the robot.

	Left Sensor	Center Sensor	Right Sensor	Wall Left	Wall Center	Wall Right	Left Motor	Right Motor	Left Signal	Right Signal	Comments
0	0	0	0	0	0	0	1	1	0	0	Forward
1	0	0	1	0	0	1	1	1	0	0	Forward
2	0	1	0	0	1	0	1	0	0	1	Turn Right
3	0	1	1	0	1	1	0	1	1	0	Turn Left
4	1	0	0	1	0	0	1	1	0	0	Forward
5	1	0	1	1	0	1	1	1	0	0	Forward
6	1	1	0	1	1	0	1	0	0	1	Turn Right
7	1	1	1	1	1	1	1	0	0	1	Reverse

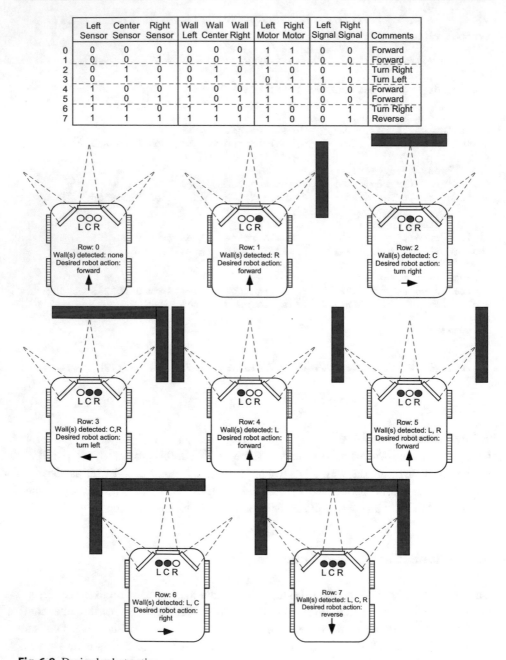

Fig. 6.8 Desired robot action

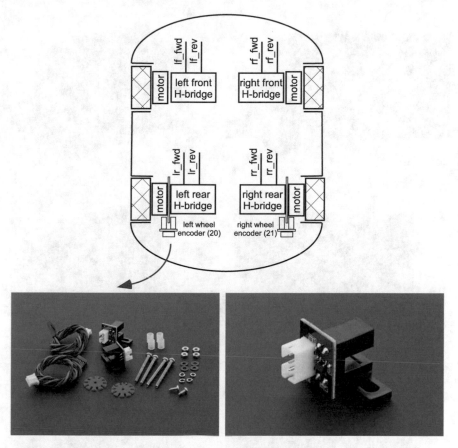

Fig. 6.9 DF Robot Gravity: TT motor encoder kit. Image used courtesy of DF Robot (www.dfrobot.com)

Figure 6.10 shows Tyler's early conceptual drawings of the Earth Roamer–inspired robot chassis rendered in the popular, free, browser–based CAD software Onshape (www.onshape.com). Note the extra space in the top rear of the chassis for additional instrumentation. Tyler aimed to create a stout design that could fit the large components of the four–wheel drive robot, while also including a window on the roof to see and potentially adjust the IR sensors, LCD screen, and LEDs as needed. The trim shown in Fig. 6.11 serves as decoration, but also as a means to link together separate pieces of the chassis, discussed below.

The size of the robot measured 15 in. from bumper to bumper and 8.5 in. wide, which is too large to fit on smaller 3D printers such as the Prusa i3 MK3S+. Instead, we split the chassis into eight smaller parts, and then 3D printed each part on larger Ultimaker FDM 3D printer. Each part was designed with interlocking pins that could be linked together once printed, and the entire model was designed to print with minimal or no support in order to minimize the time needed to print each part. In spite of all these measures, because of

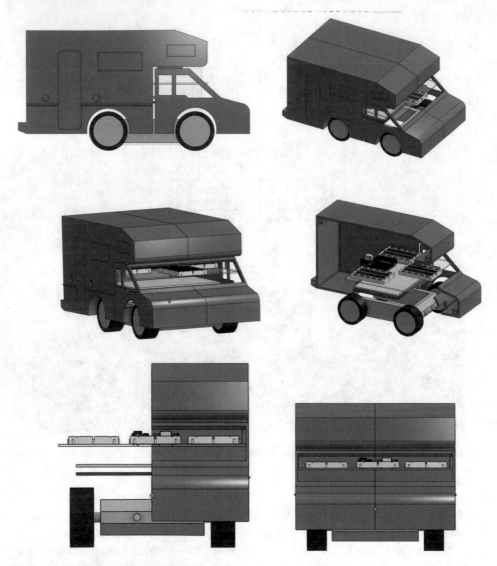

Fig. 6.10 Earth Roamer chassis concept

their size, each part took approximately 15–24 h to print! **Pop quiz:** can you see why even the pin holes on the Earth Roamer chassis in Fig. 6.12 do not require supports? Hint: it has something to do with the Y–H–T rule.

We set the infill percentage of the trim to be 5%, which you can see printing in Fig. 6.12. **Pop quiz:** why were we confident that we didn't need dense infill for the decorative trim? Likewise, because of the size of each part, we did not need to print any type of bed adhesion

Fig. 6.11 Final concept model of our Earth Roamer four–wheel drive robot, split into eight different 3D printable parts

besides a skirt. Can you recall the types of situations that would require 3D printing with a brim, and why this project did not require one?

6.2.8 Mountain Maze

The mountain maze is constructed from plywood, chicken wire, expandable foam, plaster cloth and Bondo. A rough sketch of the desired maze path is first constructed. Care is taken to ensure the pass is wide enough to accommodate the robot. The maze platform is constructed from 3/8 in. plywood on 2 by 4 in. framing material. Maze walls are also constructed from the plywood and supported with steel L brackets.

With the basic structure complete, the maze walls are covered with chicken wire. The chicken wire is secured to the plywood with staples. The chicken wire is then covered with plaster cloth (Creative Mark Artist Products #15006). To provide additional stability, expandable foam is sprayed under the chicken wire (Guardian Energy Technologies, Inc. Foam It Green 12). The mountain scene is then covered with a layer of Bondo for additional structural stability. Bondo is a two–part putty that hardens into a strong resin. Mountain pass construction steps are illustrated in Fig. 6.13. The robot is shown in the maze in Fig. 6.14.

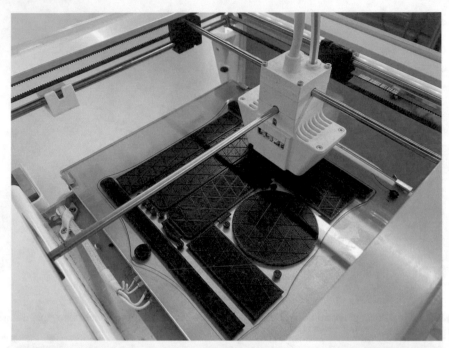

Fig. 6.12 Parts of the Earth Roamer chassis being 3D printed on a large Ultimaker 3 Extended 3D printer

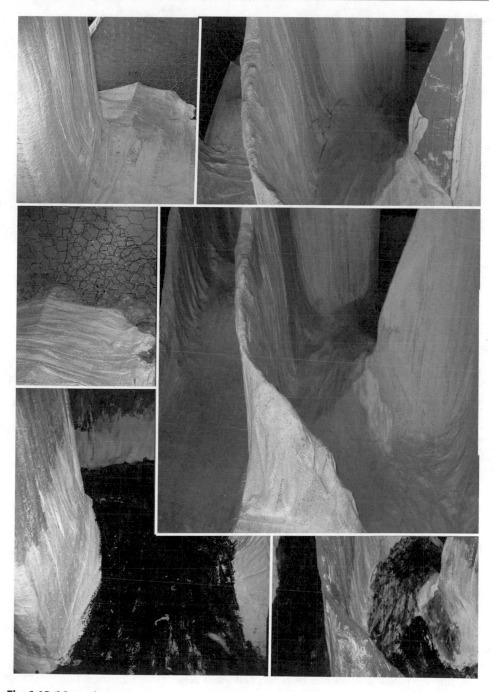

Fig. 6.13 Mountain maze

Fig. 6.14 Earth Roamer inspired robot in the mountain maze

6.2.9 Project Extensions

– Modify the turning commands such that the PWM duty cycle and the length of time the motors are on are sent in as variables to the function.
– Equip the robot wheels with encoders to control the length of a turn.
– Equip the motor with another IR sensor that looks down toward the maze floor for "land mines." A land mine consists of a paper strip placed in the maze floor that obstructs a

portion of the maze. If a land mine is detected, the robot must deactivate it by moving slowly back and forth for three seconds and flashing a large LED.

- Develop a four wheel drive system which includes a tilt sensor. The robot should increase motor RPM (duty cycle) for positive inclines and reduce motor RPM (duty cycle) for negatives inclines.
- Equip the robot with an analog inertial measurement unit (IMU) to measure vehicle tilt. Use the information provided by the IMU to optimize robot speed going up and down steep grades.
- Equip the robot with onboard sensors appropriate for a specific mission.

6.3   Summary

The purpose of this chapter was to put into practice lessons learned throughout the book. Due to space constraints we limited our exploration to only two projects. We considered a number of motivating alternatives including a robot fish, a robot snake, a bionic artificial hand, and a 4WD maze navigating robot. We ultimately chose the bionic hand because of the social impact this project may have in helping others. We also chose the 4WD maze navigating robot to demonstrate how a number of sensors and actuators may be used in a single, integrated system.

6.4   Problems

1. For the robot project, equip the robot with turn signals.
2. For the robot project, provide an audible signal when the robot is reversing.
3. For the robot project, modify the turning commands such that the PWM duty cycle and the length of time the motors are on are sent in as variables to the function.
4. For the robot project, equip the robot wheels with encoders to control the length of a turn.
5. For the robot project, equip the motor with another IR sensor that looks down toward the maze floor for "land mines." A land mine consists of a paper strip placed in the maze floor that obstructs a portion of the maze. If a land mine is detected, the robot must deactivate it by moving slowly back and forth for three seconds and flashing a large LED.
6. For the robot project, develop a four wheel drive system which includes a tilt sensor. The robot should increase motor RPM (duty cycle) for positive inclines and reduce motor RPM (duty cycle) for negatives inclines.

7. For the robot project, equip the robot with an analog inertial measurement unit (IMU) to measure vehicle tilt. Use the information provided by the IMU to optimize robot speed going up and down steep grades.
8. For the robot project, equip the robot with onboard sensors appropriate for a specific mission.

Index

© The Editor(s) (if applicable) and The Author(s), under exclusive license to Springer
Nature Switzerland AG 2022
T. Kerr and S. Barrett, *Arduino IV: DIY Robots*, Synthesis Lectures on Digital Circuits
& Systems, https://doi.org/10.1007/978-3-031-11209-6

Printed in the United States
by Baker & Taylor Publisher Services